SpringerBriefs in Electrical and Computer Engineering

Series editors

Woon-Seng Gan, Nanyang Technological University, Singapore, Singapore
C.-C. Jay Kuo, University of Southern California, Los Angeles, CA, USA
Thomas Fang Zheng, Tsinghua University, Beijing, China
Mauro Barni, University of Siena, Siena, Italy

W0235078

SpringerBriefs present concise summaries of cutting-edge research and practical applications across a wide spectrum of fields. Featuring compact volumes of 50 to 125 pages, the series covers a range of content from professional to academic. Typical topics might include: timely report of state-of-the art analytical techniques, a bridge between new research results, as published in journal articles, and a contextual literature review, a snapshot of a hot or emerging topic, an in-depth case study or clinical example and a presentation of core concepts that students must understand in order to make independent contributions.

More information about this series at http://www.springer.com/series/10059

Jiadi Yu • Yingying Chen • Xiangyu Xu

Sensing Vehicle Conditions for Detecting Driving Behaviors

 Springer

Jiadi Yu
Department of Computer Science
& Engineering
Shanghai Jiao Tong University
Shanghai, Shanghai, China

Yingying Chen
WINLAB
Rutgers University
New Brunswick, NJ, USA

Xiangyu Xu
Department of Computer Science
& Engineering
Shanghai Jiao Tong University
Shanghai, Shanghai, China

ISSN 2191-8112   ISSN 2191-8120   (electronic)
SpringerBriefs in Electrical and Computer Engineering
ISBN 978-3-319-89769-1   ISBN 978-3-319-89770-7   (eBook)
https://doi.org/10.1007/978-3-319-89770-7

Library of Congress Control Number: 2018939025

Printed on acid-free paper

This Springer imprint is published by the registered company Springer International Publishing AG part of Springer Nature.
The registered company address is: Gewerbestrasse 11, 6330 Cham, Switzerland

Preface

As more vehicles take part in the transportation system in recent years, driving or taking vehicles has become an inseparable part of our daily life. However, the increasing number of vehicles on the roads brings more traffic issues including crashes and congestions, which make it necessary to sense vehicle conditions and detect driving behaviors for both drivers and other participants in the transportation system. For example, sensing lane information of vehicles in real time can be assisted with the navigators to avoid unnecessary detours, and acquiring instant vehicle speed is desirable to many important vehicular applications, such as vehicle localization, and building intelligent transportation systems. Moreover, if the driving behaviors of drivers, such as inattentive and drunk driving, can be detected and warned in time, a large part of traffic accidents can be prevented. In addition, vehicle dynamics and driving behaviors can also be applied in the way of crowd sensing to help the traffic planers analyze the traffic conditions and make proper decisions.

For sensing vehicle dynamics and detecting driving behaviors, traditional approaches are grounded on the built-in infrastructure in vehicles, such as infrared sensors and radars, or additional hardware like EEG devices and alcohol sensors. However, currently, only the high-end vehicles are equipped with such a built-in infrastructure, and the implementation of additional devices could be less convenient and costly. With the tremendous development of mobile devices in recent years, this book provides a different approach by utilizing the built-in sensors of smartphones to sense vehicle dynamics and detect driving behaviors.

The book begins by introducing the concept of smartphone sensing, and also presenting the motivation behind the book and summarizing our contributions to the book. After that, in Chap. 2, we propose approaches utilizing built-in sensors in smartphones for sensing vehicle dynamics. In this chapter, we begin with sensor data processing including coordinate alignment and data filtering, then the sensing data are exploited to detect various vehicle dynamics, including moving, driving on uneven road, turning, changing lanes, and the instant speed of vehicles. In Chap. 3, we discuss the detection of abnormal driving behavior by sensing vehicle dynamics. Specifically, we propose a fine-grained abnormal Driving behavior Detection and iDentification system, D^3, to perform real-time high-accurate abnormal driving behaviors monitoring using the built-in motion sensors in smartphones. D^3 system is evaluated in real driving environments with multiple drivers driving for months. In Chap. 4, we exploit the feasibility to recognize abnormal driving events of drivers at early stage. In real driving environment, providing detection results after an abnormal driving behavior is finished is not sufficient for alerting the driver and avoiding car accidents. Thus, early recognition of inattentive driving is the key to improve driving safety and reduce the possibility of car accidents. Specifically, we developed an *E*arly *R*ecognition system, ER, which recognizes inattentive driving events at an early stage and provides timely alerts to drivers, leveraging built-in audio devices on smartphones. ER system is evaluated in real driving environments with multiple drivers driving for months. In Chap. 5, we provide an overview of the state-of-the-art researches. Finally, conclusions and future directions are presented in Chap. 6.

This book is supported in part by NSFC (No. 61772338, No. 61420106010) and NSF (CNS1409767, CNS1514436). Also, the authors would like to thank Prof. Xuemin (Sherman) Shen for providing the opportunity to work with Springer.

Shanghai, People's Republic of China   Jiadi Yu
New Brunswick, NJ, USA   Yingying Chen
Shanghai, People's Republic of China   Xiangyu Xu
January 15, 2018

Contents

Chapter 1
Overview

1.1 Brief Introduction of Smartphone Sensing

Although early mobile phones are designed to communicate only, modern smartphones are developed more as a ubiquitous mobile computing device, and thus become more and more central to people's lives. What makes smartphones even more powerful are the varies kinds of sensors embedded in smartphones. Sensor embedded smartphones are able to set as a platform for numerous applications, such as recreational sports, environmental monitoring, personal health-care, sensor augmented gaming, virtual reality, and smart transportation systems.

Moreover, the increasing popularity of smartphones and the rapid development of their sensing capability show great availability for researchers to exploit smartphones for sensing tasks. Meanwhile, with the environments for smartphone applications (e.g., Apple AppStore and Google Android Market) more matured in recent years, it is more convenient for researchers to extend the traditional small scale laboratory-level deployments to large scale experiments in the real-world involves a large number of users. Together, these effects bring smartphone sensing a hot research issue, where new models, algorithms, systems and applications spring up in recent years.

1.1.1 Representative Sensors Embedded in Smartphones

A smartphone can be thought as a small-size mobile computer with sensors. As computers, the computation and storage capabilities of smartphones keep growing every year. As sensing platforms, smartphones integrate varies types of sensors, including accelerometer, gyroscope, magnetometer (compass), GPS, microphone,

© The Author(s), under exclusive licence to Springer International Publishing AG, part of Springer Nature 2018
J. Yu et al., *Sensing Vehicle Conditions for Detecting Driving Behaviors*, SpringerBriefs in Electrical and Computer Engineering, https://doi.org/10.1007/978-3-319-89770-7_1

Fig. 1.1 Illustration of the
smartphone's coordinate
system for accelerometer and
gyroscope

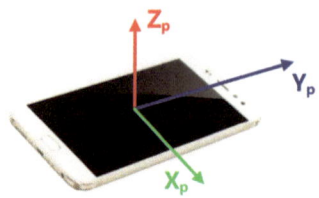

cameras, etc. These sensors embedded in a smartphone can be classified into several
categories, as shown in the following:

Motion Sensors Motion sensors in a smartphone are commonly embedded as
inertial measurement unit (IMU), which is an electronic device that measures the
accelerate, angular rate, and sometimes other motion information of an object.
Typically, IMU in a smartphone contains a 3-axis accelerometer and a 3-axis
gyroscope. Specifically, the 3-axis accelerometer measures the acceleration of the
smartphone from three axis that are orthogonal to each other, as x, y and z
axis shown in Fig. 1.1. Similarly, gyroscope measures the angular speed of the
smartphone for the rotation of these three axis.

Optical Sensors Optical sensors in a smartphone appear as cameras (commonly
a front camera and a rear camera), light sensor and proximity sensor, each of
which has different function and mechanism. Cameras give image information
based on Charge Couple Device (CCD) and other auxiliary circuits. The light sensor
measures the intensity of ambient light through the ability of light-sensitive semi-
conduct. For proximity sensor, it contains an infrared LED and an infrared light
detector. The sensor works by shinning a beam of infrared light which is reflected
from a nearby object and picked up by the detector.

Audio Sensors Audio sensors in a smartphone includes a speaker and several
microphones (commonly two microphone, one for collecting audios, one for reduce
the noise). The speaker is made to create acoustic signals, while the microphones
collect them. Both the speaker and microphones embedded in a smartphone have a
working frequency range of 15 Hz to 24 kHz, covering the frequency range that a
person can speak or hear.

Other Sensors Besides sensors mentioned above, there are still a list of other
sensors embedded in a smartphone nowadays, e.g., position sensors (including GPS
and magnetometers), environment sensors (including barometers, hygrometers and
thermometers), etc.

1.1.2   Development of Smartphone Sensing

Smartphone sensing technology has been developing ever since smartphones with
sensors appeared. However, although equipped with several useful sensors (e.g.,

camera, accelerometer, GPS, etc.), early smartphones are limited in finishing sensing tasks. These limitations were reflected in several aspects. Firstly, these smartphones lack efficient application programming interfaces (APIs) to access some of the phone's components, making it hard for researchers to create mobile sensing applications. Secondly, the storage and computation ability of early smartphones are limited by hardware, which means these smartphones cannot handle large amount of sensing data or perform complex sensing tasks. Thirdly, the battery of early smartphones are not powerful enough for continuously sensing.

Recent years, Android and iOS smartphone operating systems are much more supportive platforms for mobile sensing application programming. They provide a more complete set of APIs to access the low level components of the phone operating system (OS) while taking advantage of more powerful hardware (CPU and RAM). Although, the smartphone battery capacity is still a bottleneck, it is possible to realize continuously sensing by fine-grained resource management.

In the upcoming future, there are other factors that need to be taken into consideration in order to make a mobile sensing application successful. These are issues related to sensor readings interpretation, inference label accuracy and validation. For instance, when a machine learning algorithm is deployed to run in the wild without supervision, the lack of ground truth evidence calls for mechanisms to validate the sensing data. Also, although they have increasingly powerful hardware platforms, smartphones still have limitations in running computationally intensive algorithms, such as the learning phase of a machine learning technique. External, more powerful support, e.g., cloud computing, could be exploited to mitigate this problem.

1.2 Smartphone Sensing in Vehicles

Since driving safety has gained increasing attention during recent decades, there has been active research efforts towards developing sensing technologies in vehicles. Traditional sensing technologies in vehicles depends on specialized sensors, e.g., EEG machine, high-precision radar and camera, which involves extra cost and can only focused on a certain type of tasks. More recently, mobile devices like smartphones have been exploited to reinforce driving safety and improve driving experience. Differed from traditional sensing technologies with heavy infrastructures, with a rich set of sensors and the abilities in storage and computation, smartphones are appeared to be convenient to carry and available for varies types of sensing tasks in vehicles.

Generally, the sensing tasks in vehicles can be divided in three main categories as following:

- **Sensing Conditions Outside Vehicles**: focuses on enhancing the sensing abilities of drivers, making them more sensitive to the conditions of roads, traffic signs and other traffic participants. As the roads become more complicated and traffic participants increases greatly during recent decades, drivers are required to sense more information in order to make themselves safe and comfortable while driving, which requires drivers to be more focused and, as a result, makes drivers easy to be exhausted. By exploiting the sensing ability of smartphones, e.g., using cameras in smartphones to recognized other vehicles and traffic signs, it is possible for drivers to spend less effort and sense more traffic conditions.

- **Sensing Conditions of Vehicles**: aims at monitoring the conditions of vehicles, including current speed, location, etc., and behaviors of vehicles, such as turning, changing lanes, accelerating and breaking. As the basic unit to construct Intelligent Transportation System (ITS), it is essential to capture the conditions and behaviors of vehicles. However, even to nowadays, only some of the high-end vehicles are equipped with enough sensors for monitoring vehicle conditions, making it necessary to explore smartphones to sensing conditions of vehicles. Typically, the motion sensors in a smartphone (accelerator and gyroscope) are appropriate to detect the vehicle conditions.
- **Sensing Behaviors of Drivers**: studies on detecting different driving behaviors of drivers when they are driving, especially the behaviors that are dangerous, including inattentive, distracted, fatigue, drunk, using mobile devices while driving, etc. Due to the independence of each vehicle in the road, these behaviors of drivers are hard to be detected in time from outside the vehicles, and sometimes even difficult to be noticed by the drivers themselves. As a result, to prevent dangerous driving behaviors by monitoring the behaviors of drivers, smartphones are exploited to be the detector.

1.3 Overview of the Book

In this book, we elaborate the studies on sensing vehicle conditions and driving behaviors with smartphones based on realistic data collected in real driving environments. The reminder of this book is organized as follows:

In Chap. 2, we present our works on leveraging the built-in motion sensors of smartphones to detect various vehicle conditions, such as stopping, turning, changing lanes, driving on uneven road, etc. By analyzing the data of accelerometer and gyroscope collected from real driving environments, we find that different vehicle conditions are reflected as specific patterns on sensor readings. Based on this observation, we further propose several approaches to process the motion sensor readings of smartphone for sensing different types of vehicle conditions. By performing coordinate alignment and data filtering, we first relate the motion sensor readings of smartphone to the motion of vehicle. After that, motion sensor readings are exploited to find the pattern for different vehicle conditions, including stopping, turning, changing lanes and driving on uneven road, and we further estimate the instant speed of vehicle based on these vehicle conditions. From results of extensive experiments involving five volunteers driving for months, we show the feasibility of the proposed approaches for sensing vehicle conditions in real driving environments.

In Chap. 3, we describe a fine-grained monitoring approach for abnormal driving behaviors of vehicles, which not only detects abnormal driving behaviors in real-time, but also identifies specific types of abnormal driving behaviors, i.e. *Weaving, Swerving, Side-slipping, Fast U-turn, Turning with a wide radius* and *Sudden braking*. Through empirical studies of the 6-month driving traces collected from real driving environments, we find that all of the six types of driving behaviors have their unique patterns on acceleration and orientation. Recognizing this observation, we further propose a fine-grained abnormal *D*riving behavior *D*etection and i*D*entification system, D^3, to perform real-time high-accurate abnormal driving behaviors monitoring using smartphone sensors. By extracting unique features from readings of smartphones i.e., accelerometer and orientation sensor, we first identify 16 representative features to capture the patterns of driving behaviors. Then, a machine learning method, Support Vector Machine (SVM), is employed to train the features and output a classifier model which conducts fine-grained identification. From results of extensive experiments with 20 volunteers driving for another 4 months, we show the feasibility of D^3 system in real driving environments.

In Chap. 4, we present an approach that leverage built-in audio devices on smartphones to realize early recognition of inattentive driving events including *Fetching Forward, Picking up Drops, Turning Back* and *Eating or Drinking*. Through empirical studies of driving traces collected in real driving environments, we find that each type of inattentive driving event exhibits unique patterns on Doppler profiles of audio signals. This enables us to develop an *E*arly *R*ecognition system, ER, which can recognize inattentive driving events at an early stage and alert drivers timely. ER employs machine learning methods to first generate binary classifiers for every pair of inattentive driving events, and then develops a modified vote mechanism to form a multi-classifier for all four types of inattentive driving events, for atypical inattentive driving events along with other driving behaviors. It next turns the multi-classifier into a gradient model forest to achieve early recognition of inattentive driving. Through extensive experiments with eight volunteers driving for about 2 months, ER can achieve high accuracy for inattentive driving recognition and realize recognizing events at very early stage.

In Chap. 5, we give a brief review of state-of-art works related to the approaches presented in this book. This chapter starts with introducing the representative researches of smartphone sensing. Then researches focusing on vehicle dynamics sensing are mentioned. After that, we gives introduction to driver behaviors detection researches in recent decades. Finally, we list the common issues of these state-of-art researches.

In Chap. 6, we show that works presented in this book identifies new problems and solutions of leveraging smartphone to sense vehicle conditions and detect driving behaviors, which helps us to advance our understanding in smartphone sensing and its applications related to driving. We present the future direction of our research at the end of the book.

Chapter 2
Sensing Vehicle Dynamics with Smartphones

2.1 Introduction

The smartphone-based vehicular applications become more and more popular to analyze the increasingly complex urban traffic flows and facilitate more intelligent driving experiences including vehicle localization [1, 2], enhancing driving safety [3, 4], driving behavior analysis [5, 6] and building intelligent transportation systems [7, 8]. Among these applications, the vehicle dynamics is an essential input. Accurate vehicle dynamic detection could make those vehicle-dynamic dependent applications more reliable under complex traffic systems in urban environments.

When driven, vehicles can be considered as moving objects with their own dynamics, including accelerate, angular speed, etc. These dynamics, which change over time, show the basic moving characters of vehicles, such as turning, changing lanes, braking, etc. Typically, some of these dynamics are available from the vehicle system, e.g., the speed of vehicle can be obtained from CAN bus of vehicle by using an additional OBD adapter, while other dynamics, such as turning and changing lanes, are not measured by the vehicle system. Since drivers usually take their smartphones when driving, it is reasonable to utilize sensors in smartphones to sense the dynamics of vehicles. The motion sensors embedded in driver's smartphone can detect the moving dynamics of the smartphone, which is highly related to the dynamics of vehicle. As a result, we propose new approaches to sense vehicle dynamics with smartphones for developing useful smartphone-based vehicular applications.

In order to successfully sensing the dynamics of vehicles with smartphones, we could leverage the sensors embedded in smartphones. Among all kinds of sensors, the motion sensors in smartphones (accelerometer and gyroscope) are most suitable for this task, as they directly measure the physical parameters for motions of smartphones. Moving along this direction, there are several challenges to be solve.

J. Yu et al., *Sensing Vehicle Conditions for Detecting Driving Behaviors*,
SpringerBriefs in Electrical and Computer Engineering,
https://doi.org/10.1007/978-3-319-89770-7_2

Firstly, the coordinate system of smartphone is different from the coordinate system of vehicle, so the sensor readings that measures the smartphone motions need to be transferred into vehicle dynamics. Secondly, the environment of real driving environments is full of noises, which need to be filtered. Thirdly, effective features of different vehicle behaviors need to be extracted from sensor readings. Finally, the solutions need to be light-weight for implementation in common smartphones.

In this chapter, we focus on leveraging the accelerometer and gyroscope for sensing vehicle dynamics. To achieve the sensing tasks, sensor readings first need to be pre-processed before further utilized. Specifically, there are two methods: *Coordinate Alignment* is applied to align the smartphone's coordinate system with the vehicle's, while *Data Filtering* is introduced to remove the noises in sensor readings. After pre-processing, we show that different types of vehicle behaviors, such as stopping, turning, changing lanes, can be detected by a smartphone through the proposed approaches, and we further propose an approach to accurately estimate the instant speed of vehicles based on sensing these vehicle behaviors. Our extensive experiments validate the accuracy and the feasibility of proposed approaches in real driving environments.

The rest of the chapter is organized as follows. The pre-processing of sensor readings is presented in Sect. 2.2. Section 2.3 shows the approaches for measuring different basic vehicle dynamics. We evaluate the performance of these approaches and present the results in Sect. 2.4. Finally, we give our solution remarks in Sect. 2.5.

2.2 Pre-processing Sensor Readings

2.2.1 Coordinate Alignment

We cannot derive meaningful vehicle dynamics from sensor readings on the smartphone unless the phone's coordinate system is aligned with the vehicle's. The Coordinate Alignment sub-task aligns the phone's coordinate system with the vehicle's by utilizing the accelerometers and gyroscopes on smartphones. As illustrated in Fig. 2.1, the phone's coordinate system ($\{X_p, Y_p, Z_p\}$) is determined by the pose of the phone. Our coordinate alignment aims to find a rotation matrix R to rotate the phone's coordinate system to match with the vehicle's ($\{X_v, Y_v, Z_v\}$).

Fig. 2.1 Illustration of the vehicle's coordinate system and the smartphone's coordinate system

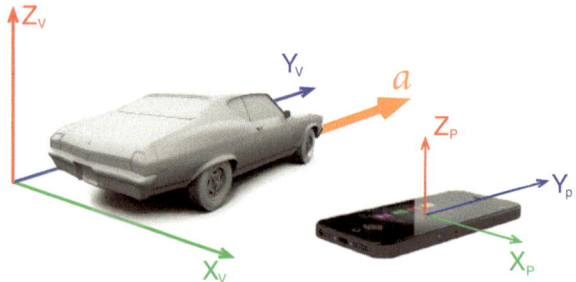

We define three unit coordinate vectors under the vehicle's coordinate system as $\hat{i}, \hat{j}$ and $\hat{k}$ for X_v, Y_v and Z_v axis respectively (i.e., $\hat{i} = [1, 0, 0]^T$ in vehicle's coordinate system). We denote the corresponding coordinates of these three unit vectors in the phone's coordinate system as:

$$\hat{q} = [x_q, y_q, z_q]^T, \tag{2.1}$$

where $q \in i, j, k$, and the rotation matrix is given by:

$$R = \left\{ \begin{matrix} x_i & x_j & x_k \\ y_i & y_j & y_k \\ z_i & z_j & z_k \end{matrix} \right\}. \tag{2.2}$$

Our coordinate alignment sub-task utilizing smartphone's accelerometer and gyroscope readings to obtain each element in the rotation matrix R consists of three steps:

Deriving $\hat{k}$ We can apply a low pass filter (e.g., exponential smoothing) on the three axes accelerometer readings on the phone to obtain the constant components from these three accelerations and derive the gravity acceleration [1], which is then be normalized to generate the unit vector $\hat{k} = [x_k, y_k.z_k]^T$.

Deriving $\hat{j}$ To obtain $\hat{j}$, we utilize the fact that the three-axes accelerometer readings of the phone are caused by vehicle's acceleration or deceleration when driving straight. For example, we can obtain $\hat{j} = [x_j, y_j.z_j]^T$ through extracting the accelerometer readings when the vehicle decelerates (e.g., the vehicle usually decelerates before making turns or encountering traffic lights and stop sign). The gyroscope is used to determine whether the vehicle is driving straight (i.e., with zero rotation rate). We note the gravity component needs to be excluded because it distributes on all three axes of the phone when the phone's coordinate system is not aligned with the vehicle's.

Obtaining $\hat{j}$ Since the coordinate system follows the right hand rule, we can determine the unit vector $\hat{i} = \hat{j} \times \hat{k} = [x_i, y_i, z_i]^T$.

Recalibrating $\hat{k}$ Consider the scenario that the gravity direction does not align with the z-axis of the vehicle when the vehicle is running on a slope. In order to keep the orthogonality of the vectors in rotation matrix, we recalibrate the z-axis vector by $\hat{k} = \hat{i} \times \hat{j}$.

After obtaining the rotation matrix R, given the sensor reading vector in the phone's coordinate system s, we can obtain the rotated sensor reading vector s' aligned with vehicle's coordinate system by applying a rotation matrix R as: $s' = s \times R$. We note that there are existing studies utilizing the sensors embedded in smartphones to calibrate the coordinate systems between the phone and the vehicle.

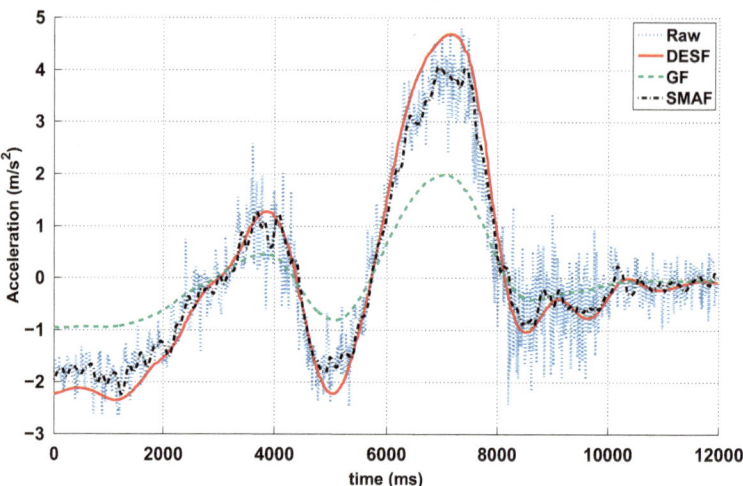

Fig. 2.2 Raw acceleration and the resulted acceleration after applying SMAF, DESF and GF

Different from the previous study, our coordinate alignment mechanism does not require working with the GPS on the phone, and thus is more accurate and energy efficient.

2.2.2 Data Filtering

Due to environmental dynamics, there are many high frequency noises in the collected raw data of motion sensors, e.g., the vibrate of vehicle caused by uneven road and the small movement of smartphones during the driving, which may have negative impacts on analyzing the dynamics of vehicle behaviors. Thus we conduct data filtering to the collected raw data. Specifically, low-pass filter is utilized to remove the high frequency noise and yet capture the statistical features presented in the traces.

In practice, there are various common low-pass filters, each of which has different performances on different types of data. We experiment with three most common low-pass filters in our dataset, i.e. simple moving average filter (SMAF), dynamic exponential smoothing filter (DESF) and Gaussian filter (GF) [9], and compare their performances to finally select one filter that efficiently eliminate high frequency noises while preserving the effective dynamics of different vehicle behaviors to the utmost extent.

Figure 2.2 shows an example including the raw acceleration readings from smartphones' accelerometer and the resulted accelerations after applying each of the three low-pass filters. It can be seen from Fig. 2.2 that SMAF keeps the most changes in data, but it cannot perfectly remove high frequency noises, while GF gets the most smooth curve by eliminating high frequency noises, but it cannot preserve effective features of vehicle behaviors. Compared with the rough trend of the acceleration preserved by SMAF and GF, DESF is able to not only return the nicely fitted curve but also preserve effective features of different vehicle behaviors, which is the best filter for our tasks among the three filters. This is because DESF is an exponential smoother that changes its smoothing factor dynamically according to previous samples, making it suitable for processing dynamic data. Hence, we select DESF as the low-pass filter to remove high frequency noises as well as preserve the effective dynamics of vehicle behaviors to the utmost extent.

2.3 Sensing Basic Vehicle Dynamics

2.3.1 Sensing Movement of Vehicles

When driving, the movements of vehicle can be roughly divided into two states: moving state and stopping state. Typically, these two states appears alternately during a driving, e.g., a driver may stops the vehicle at the red traffic light after moving for some time, and may starts again for moving until encountering next red light. Compared to moving state, stopping state lasts for much shorter time and thus can be detected as a special type of vehicle dynamic. If stopping state can be detected by a smartphone, then the full picture of moving states and stopping states during driving can be obtained from the smartphone.

Based on our analysis on sensor readings, the data pattern of the acceleration on the vehicle's z-axis for stopping state is remarkably different from that of moving state. Figure 2.3 plots the readings from the accelerometer's z-axis when the vehicle

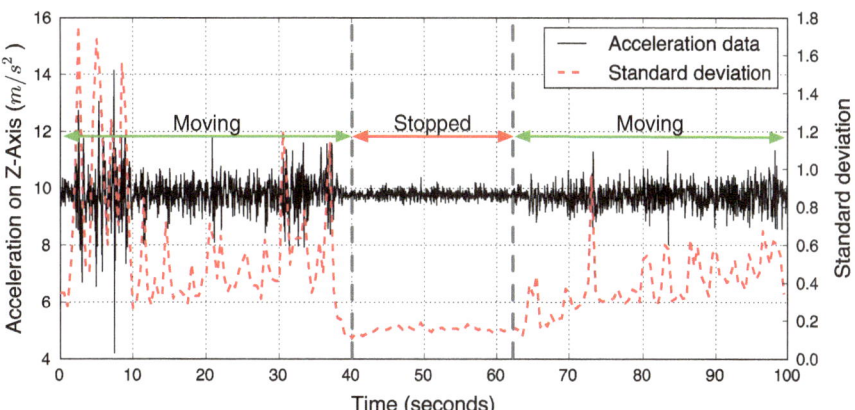

Fig. 2.3 Illustration of the acceleration on the vehicle's z-axis and the corresponding standard deviation when a vehicle stops

is in moving state and stopping state. From Fig. 2.3, it can be seen that the jitter of the acceleration on z-axis is almost disappeared and the standard deviation of the acceleration on z-axis remains low while the vehicle stops. The standard deviation of the acceleration collected by the smartphone is calculated in a small sliding window (usually the size of the window is set as 1 s, which is determined through empirical studies).

2.3.2 Sensing Driving on Uneven Road

Besides driving on the plain road with little impact on sensor readings, when driving on uneven road conditions, sensor readings in a smartphone may be interfered. Since speed bumps, potholes, and uneven road surfaces are common in real traffic environments, it is necessary to detect when vehicles are driving on uneven road for further analyzing sensor readings in smartphones.

Specifically, when a car is passing over uneven road surfaces, the acceleration on the vehicle's z-axis shows particular patterns. Figure 2.4 shows the accelerations on the car's z-axis, when a car is passing over a speed bump. The front wheels hit the bump first and then the rear wheels. In Fig. 2.4, the first peak is produced when the front wheel is passing over the bump and the second peak is produced by the rear wheels. It can be seen from the figure that the lasting time interval ΔT is key to observe the impact of uneven road conditions on sensor readings.

Considering the similarity between the two peaks, we use *auto-correlation* analysis to find ΔT. Given an acceleration sequence on z-axis, $\{Acc\}$, auto-correlation of lag τ is:

$$R(\tau) = \frac{E[(Acc_i - \mu)(Acc_{i+\tau} - \mu)]}{\sigma^2}, \tag{2.3}$$

where μ is the mean value of Acc and σ is the standard deviation. Figure 2.4 also shows the auto-correlation results of the accelerometer's readings on z-axis. Obviously, $R(\tau)$ is an even function, so $R(\tau) = R(-\tau)$. To get the ΔT, we need to find the maximum peak value except the one at $\tau = 0$, and the horizontal distance from the maximum peak to $\tau = 0$ equals to ΔT. The time $\tau = 0$ can be obtained by checking the value of Acc and when it is larger than a threshold which can be learned through empirical studies, such a time point is regarded as R(0). And for the wheelbase, we can get it from vehicle's product specifications.

Additionally, the phone's location in the car has an influence on the shape of the two acceleration peaks. Specifically, if the phone is located in the front row, then the first peak has a larger amplitude than that of the second peak and vice versa. However, in our experiment, the phone's location is not changed during driving. Thus whatever the shapes of the two acceleration peaks, they are always similar to each other and can be detected by auto-correlation. Therefore, the phone positions will not affect the detection of ΔT.

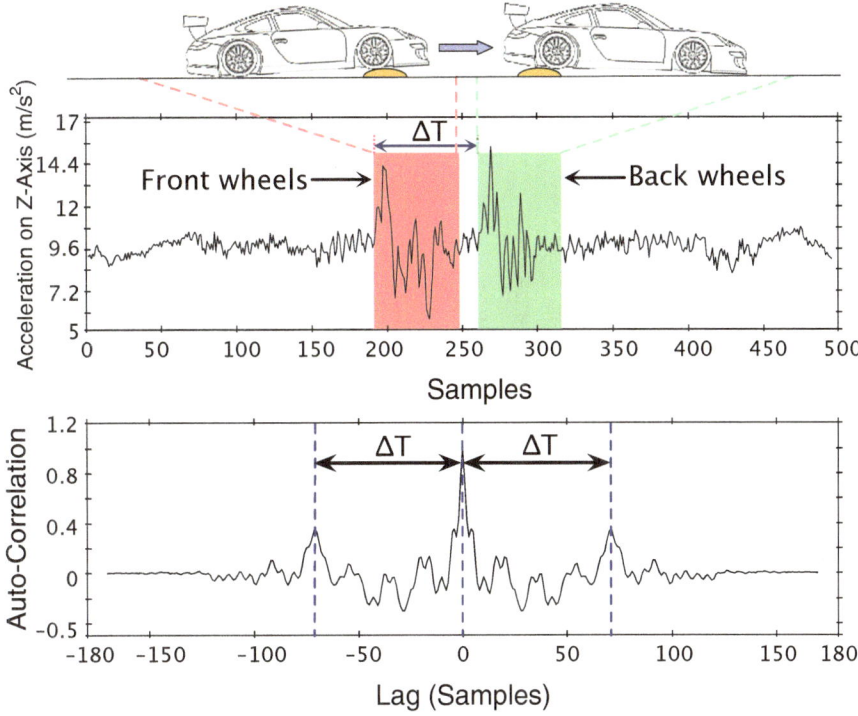

Fig. 2.4 Illustration of the acceleration on the vehicle's *z*-axis and the corresponding auto-correlation results when a car is passing over a bump

2.3.3   Sensing Turning of Vehicles

Turning is a common vehicle dynamic with great significance, as it involves changes of the moving directions of vehicles. Moreover, analyzing the vehicle dynamics during turns may be helpful to detect abnormal vehicle behaviors like over-steer and under-steer. Therefore, it is necessary to detect whether a vehicle is turning or not through smartphones.

Specifically, when a vehicle makes a turn, it experiences a centripetal force, which is related to its speed, angular speed and turning radius. Thus, by utilizing the accelerometer and the gyroscope, we can derive the tangential speed of a vehicle. Suppose a car is turning right, as is shown in Fig. 2.5, then $v = \omega R$, $a = \omega^2 R$, and $\omega = \omega'$, where a is the centripetal acceleration, ω' is the angular speed of the car, R is the turning radius and ω is the angular speed that is related to the center of the orbit circle. Thus, we obtain

$$v = \frac{a}{\omega'}. \tag{2.4}$$

Fig. 2.5 Illustration of the circular movement when a car makes a turn

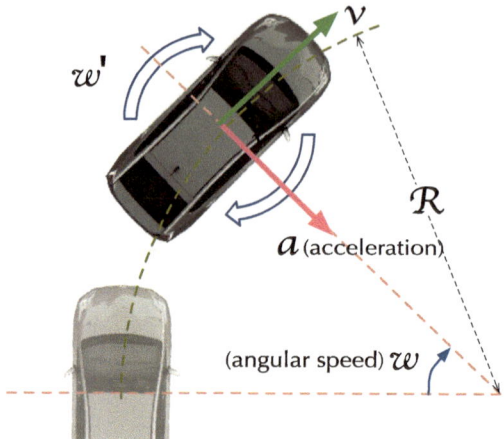

Since the centripetal acceleration a and the angular speed ω can be obtained from the accelerometer and the gyroscope respectively, the speed can be calculated based on Eq. (2.4).

2.3.4 Sensing Lane-Changes of Vehicles

The improving of transportation system in recent decades makes road much wider than it used to be, which makes lane-change an inevitable part of driving. Since it is useful in practice to detect lane-change for vehicle to help with driving. We propose an practical approach for sensing lane-changes of vehicles [10, 11]. Typically, there are two types of lane-changes in real driving environments, i.e., *single lane-changes* and *sequential lane-changes*, each of which has different directions, as shown in Fig. 2.6.

2.3.4.1 Identifying Single Lane-Change

Single lane-changes, as shown in Fig. 2.6a, b, are lane-changes are either leftward or rightward and are not limited to the neighbor lane. From our experiments in real driving environments, it can be seen that the acceleration pattern of the vehicle for a rightward lane-change appears as the up-to-down sinusoid in Fig. 2.7a. Similarly, the down-to-up sinusoid in Fig. 2.7b represents a leftward single lane-change.

Specifically, we applies a trained threshold δ to detect single lane-changes. If the peak of a semi-wave is larger than δ or the valley is less than $-\delta$, the semi-wave is then considered as *eligible*. In real driving environments, L^3 retrieves acceleration from the accelerometer in a smartphone in real time and then compares the acceleration with the threshold δ to determine whether a lane-change happens.

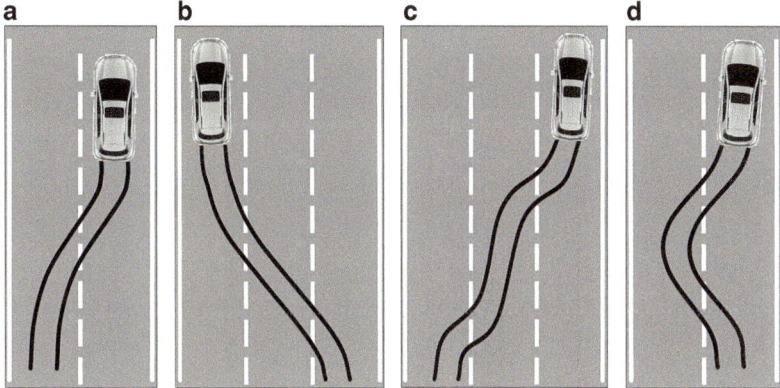

Fig. 2.6 Illustration of different lane-change behaviors

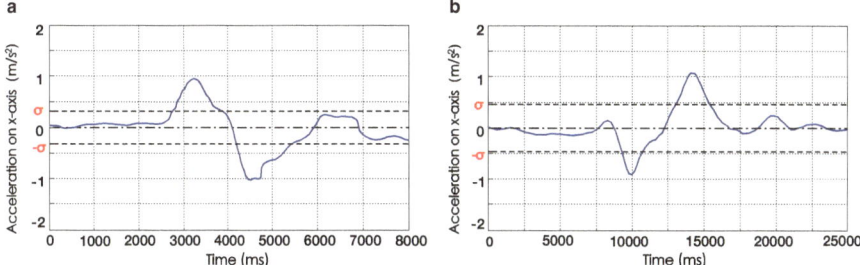

Fig. 2.7 The lateral acceleration of single lane-changes for a vehicle. (**a**) Rightward single lane-change. (**b**) Leftward single lane-change

As shown in Fig. 2.7, if there is acceleration greater/less than the threshold $\delta/-\delta$, it is considered as the peak/valley of an eligible semi-wave which starts at t_1 and ends at t_2. Then the system looks for a corresponding eligible semi-wave with valley/peak between t_2 and $t_2 + (t_2 - t_1)$. If it is found, the whole sinusoid is taken as the pattern of a rightward/leftward single lane-change.

2.3.4.2   Identifying Sequential Lane-Change

Sequential lane-changes, as shown in Fig. 2.6c, d, are the combination of several single lane-changes. If these lane-changes are independent to each other, then there are no overlaps between their waveforms of acceleration. So they can be treated as multiple single lane-changes. But if these single lane-changes have influences on each other, the overlaps between the waveforms makes them more complexed than single lane-changes. As shown in Fig. 2.8, sequential lane-changes can be classified as *Syntropy Sequential Lane-Change* and *Reversed Sequential Lane-Change*.

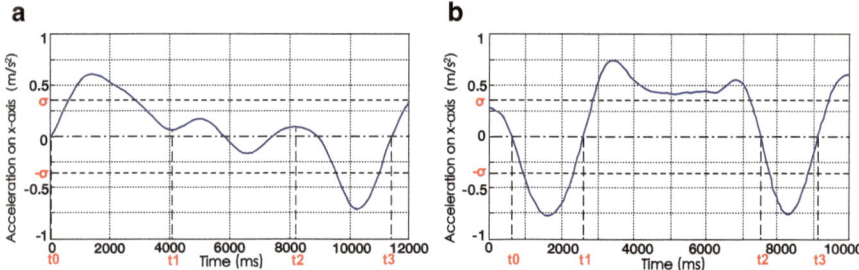

Fig. 2.8 The lateral acceleration of sequential lane-changes for a vehicle. (**a**) Syntropy sequential lane-change (rightward to rightward). (**b**) Entypy sequential lane-change (leftward to rightward)

Concretely, in **Syntropy Sequential Lane-Change**, several single lane-changes happen in the same direction. A syntropy sequential lane-change consisting of two rightward lane changes is shown in Fig. 2.6c and its waveform of lateral acceleration is shown in Fig. 2.8a. Since the connection between the two lane-changes is tight, the second semi-wave of the first lane-change counteracts with the first semi-wave of the second lane-change. Thus, the whole wave of lateral acceleration appears as two semi-waves separated by a small time span. We thus extend the *Single Lane-Change Detection* method to do the detection. After the system detects an eligible semi-wave which starts at t_0 and ends at t_1, it looks for the next eligible semi-wave with peak or valley between t_1 and $t_1 + (t_1 - t_0)$. If it is found, there is a single lane-change. If not, the system continues to look for an eligible semi-wave with peak or valley between t_2 ($t_1 \leq t_2 \leq 2t_1 - t_0$) and $t_2 + (t_1 - t_0)$. If it is found, there is a syntropy sequential lane-change. If not, the first semi-wave is taken as noise.

Similarly, in **Entypy Sequential Lane-Change**, several lane-changes happen in the reverse direction. An entypy sequential lane-change which consists of a leftward and a rightward lane-changes is shown in Fig. 2.6d and its waveform of lateral acceleration is shown in Fig. 2.8b. Since the connection between the two lane-changes is tight, the second semi-wave of the first lane-change overlaps additively with the first semi-wave of the second lane-change. Thus, the whole waveform of lateral acceleration appears as a down-to-up-to-down sinusoid. But the system cannot simply take this pattern as the waveform of entypy sequential lane-change because the last semi-wave may be a part of another lane-change. So we take advantage of another character of the entypy lane-changes.

As we know, the integral of the lateral acceleration of the vehicle is the lateral velocity difference. Therefore, for a single lane-change, the lateral velocities of the vehicle are both zero when it is about to start lane-changes and when it has just finished one. Thus, the area of the two semi-waves should be equal except a small difference due to noise. The condition can be represented as:

$$\int_{t_0}^{t_1} a(t)dt \approx - \int_{t_1}^{t_2} a(t)dt, \qquad (2.5)$$

where $a(t)$ is the lateral acceleration at time t, t_0 is the beginning time of the first semi-wave, t_1 is both the ending time of the first semi-wave and the beginning time of the second semi-wave, and t_2 is the ending time of the second semi-wave.

In the case of an entypy sequential lane-change, the area of the middle semi-wave is approximately equal to the sum of the area of the first and the third semi-waves. That is:

$$\int_{t_0}^{t_1} a(t)dt + \int_{t_2}^{t_3} a(t)dt \approx - \int_{t_1}^{t_2} a(t)dt, \tag{2.6}$$

where the first semi-wave starts at t_0 and ends at t_1, the middle semi-wave starts at t_1 and ends at t_2, the third semi-wave starts at t_2 and ends at t_3, as in Fig. 2.8b.

When the system finds an up-to-down-to-up or down-to-up-to-down sinusoid that satisfies Eq. (2.6), it takes the sinusoid as an entypy sequential lane-change.

Furthermore, the conditions of three or more lane-changes have not been considered. However, in the data we collected, three or more lane-changes appear loosely so that they can be treated as separated single lane-changes and sequential lane-changes.

2.3.5 Estimating Instant Speed of Vehicles

Among different kinds of vehicle dynamics, the instance speed of vehicle seems to be easy to obtain from the GPS embedded in vehicles or smartphones. However, GPS-based speed estimation for vehicles is not accurate enough as an input for varies kinds of vehicle-speed dependent applications. Since accurate vehicle speed estimation could make those applications more reliable under complex traffic systems in real driving environments, we develop an approach for accurately measuring the instance speed of vehicles with smartphones [12, 13].

According to our analysis in Sect. 2.2, the vehicle's acceleration can be obtained from the accelerometer sensor in the smartphone when a phone is aligned with the vehicle. Suppose the accelerometer's y-axis is along the moving direction of the vehicle, we could then monitor the vehicle acceleration by retrieving readings from the accelerometer's y-axis. The vehicle speed can then be calculated from the integral of the acceleration data over time:

$$Speed(T) = Speed(0) + \int_0^T acc(t)\, dt, \tag{2.7}$$

where $Speed(T)$ is the vehicle speed at time T and $acc(t)$ is the vehicle acceleration function of each time instant t.

Instead of producing a continuous function $acc(t)$, the accelerometer in practice takes a series of the vehicle acceleration samples at a certain sampling rate. Thus the vehicle speed can be transformed as

$$Speed(T) = Speed(0) + \sum_{i=0}^{T*k} \frac{1}{k} acc_y(i), \tag{2.8}$$

where k is the sample rate of the accelerometer and $acc_y(i)$ is the ith sample, i.e. the ith received reading from the accelerometer's y-axis. Therefore, in order to obtain the vehicle speed, we take a series of the acceleration samples by monitoring the accelerometer continuously.

Although the basic idea of using smartphone sensors to estimate vehicle speed is simple, it is challenging to achieve high-accuracy speed estimations. The most obvious problem is that sensor readings are affected by various noise encountered while driving such as engine vibrations, white noise, etc., so the estimation errors are accumulated when integrating the accelerometer's readings over time. If we can derive techniques to measure the acceleration error, the integral value of the accelerometer's readings can be corrected to get close to the true vehicle speed.

In order to measure the acceleration error, we leverage special locations in the roads called *reference point*, where the true speed is known to smartphones. Typically, there are three types of reference points, which are stopping, making turns and passing over uneven roads, as presented in Sects. 2.3.1–2.3.3, respectively. In other words, once a vehicle makes turns, stops or passes over uneven road surfaces, we are able to estimate the instant vehicle speed.

Specifically, when a vehicle stops, the instance speed of the vehicle has to be zero. In the case of making turns, the speed of vehicle can be calculated through Eq. (2.4), based on the accelerometer and angular speed captured by motion sensors on smartphones. For passing over uneven road, since we already get the time interval $\triangle T$ between these two peaks in Sect. 2.3.3, as well as the wheelbase W of the vehicle, then the vehicle speed can be measured as $v = \frac{W}{\triangle T}$.

In Fig. 2.9, the vehicle starts with zero speed, and there are two reference points P_A and P_B (i.e., the vehicle passes the reference point A and B at time T_a and T_b respectively). Suppose the integral value of the accelerometer's readings from zero to time t is $S(t)$ and the measured speed at the reference point x is RPS_x, the errors of the vehicle speed at the reference point a and b are $\triangle S(T_a) = S(T_a) - RPS_a$ and $\triangle S(T_b) = S(T_b) - RPS_b$ respectively. Since the value of acceleration error is nearly a steady constant and strongly related to the slope of the $\triangle S(t)$ curve, the acceleration error between P_A and P_B can be calculated as:

$$\tilde{A} = \frac{\triangle S(T_b) - \triangle S(T_a)}{\triangle T_a^b}. \tag{2.9}$$

where $\triangle T_a^b$ is the interval time between the reference points A and B. Thus, the accumulative error from T_a to t is $\int_{T_a}^t \tilde{A} \, dt$, i.e., $\tilde{A} \times (t - T_a)$. Furthermore, the corrected speed estimation $S'(t)$ between A and B is:

$$S'(t) = S(t) - \triangle S(T_a) - \tilde{A} \times (t - T_a). \tag{2.10}$$

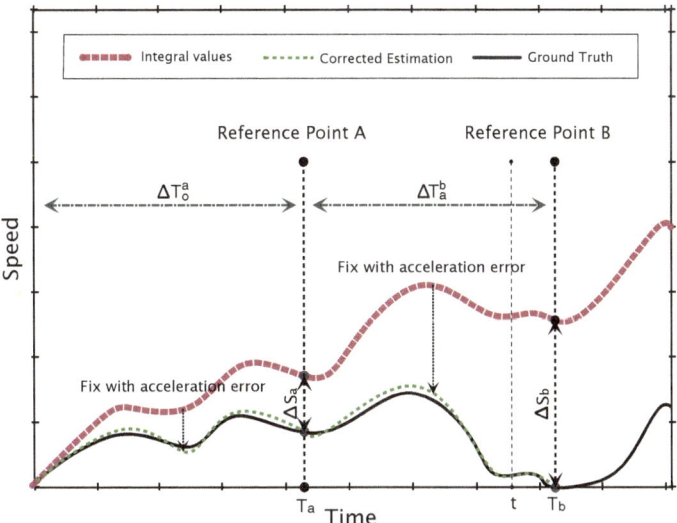

Fig. 2.9 Illustration of the acceleration error measurement using reference points

Note that the above algorithm uses the information of two adjacent reference points to correct the speed estimations between these two points. However, it is an *offline algorithm* that cannot be used for real-time speed estimations, because the information about the next reference point is unknown on real-time speed estimations. In order to achieve a real-time speed estimation, an *online algorithm* is proposed to estimate the current acceleration error. Since we know that the acceleration error changes slightly over time, thus the current acceleration error can be derived from the recent reference points. In particular, we utilize the *exponential moving average* to estimate the current acceleration error by using the recent reference points. When the ith reference point is sensed, the current acceleration error $\tilde{A}_i$ between the ith and $(i+1)^{th}$ reference point is updated through:

$$\tilde{A}_i = \alpha \cdot \tilde{A}_{i-1} + (1 - \alpha) \times \frac{\Delta S(T_i) - \Delta S(T_{i-1})}{\Delta T_{i-1}^i}, \qquad (2.11)$$

where α is the weight coefficient. α is set to be 0.5. α can determine which of the two portions is more important. The real-time speed estimation between the ith and the $(i+1)^{th}$ reference point is corrected by:

$$S'(t) = S(t) - \Delta S(T_i) - \tilde{A}_{i+1} \times (t - T_i). \qquad (2.12)$$

2.4 Evaluation

In this section, we evaluate the performance of sensing vehicle dynamics with smartphones in real driving environments.

2.4.1 Setup

We implement the approaches mentioned in Sect. 2.3 as Android Apps and install it on several smartphones, which are Samsung Galaxy Nexus, LG Nexus4, Huawei Honor3C, ZTE U809 and HTC sprint, one device for each type. Then these Apps are run by five drivers with distinct vehicles for 2-month in real driving environments to collect sensor readings for evaluation. During the 2-month, the five drivers keeps collecting data in their daily driving, including commute to work, touring, etc. Those five drivers live in different communities and they have different commute routes. Moreover, they are not told about our purpose so they can drive in a normal way. Meanwhile, we leverages vehicle DVRs to record vehicle dynamics and further asks 8 experienced drivers to recognize the vehicle dynamics as ground truth. In addition, the ground truth for speed estimation is achieved through OBD-II adapter of the vehicles.

2.4.2 Metrics

To evaluate the performance of the approaches mentioned in Sect. 2.3, let p_{ij} denote that the type of i been recognized as type of j, we define metrics as follows.

- *Accuracy*: The probability that an event is correctly identified for all K types of events, i.e., $Accuracy = \frac{\sum_{i=1}^{K} p_{ii}}{\sum_{j=1}^{K} \sum_{i=1}^{K} p_{ij}}$.
- *Precision*: The probability that the identification for an event A is exactly A in ground truth, i.e., $Precision_k = \frac{p_{kk}}{\sum_{i=1}^{K} p_{ik}}$.
- *Recall*: The probability that an event A in ground truth is identified as A, i.e., $Recall_k = \frac{p_{kk}}{\sum_{j=1}^{K} p_{kj}}$.
- *F-score*: A metric that combines precision and recall, i.e., $F\text{-}score_k = 2 \times \frac{Precision \times Recall}{Precision + Recall}$. F-score is a metric that combines precision and recall, which can provide a more convincing result than precision or recall alone.
- *False Positive Rate (FPR)*: The probability that an event not of type A is identified as A, i.e., $FPR_k = \frac{\sum_{j=1}^{K} p_{kj} - p_{kk}}{\sum_{j=1}^{K} \sum_{i=1}^{K} p_{ij} - \sum_{i=1}^{K} p_{ik}}$.
- *Average Speed Estimation Error*: The average estimation error for vehicle speed during a period of time, i.e., $Error = \frac{1}{N} \sum_{i=1}^{N} |SpeedTrue_i - SpeedEsti_i|$, where N shows the samples in the time period.

2.4.3 Performance of Sensing Vehicle Dynamics

According to our experiments, the overall accuracy for sensing stopping, turning and driving on uneven road for vehicles are 99.13%, 98.37% and 96.75% for sensing stopping, turning and driving on uneven road for vehicles, respectively, corresponding to the proposed approaches in Sects. 2.3.1–2.3.3. Further, Fig. 2.10a plots the precision, recall and F-score for sensing stopping, turning and driving on uneven road for vehicles. It can be seen that for each of these vehicle dynamics, the precision is no less than 96%, while the recall is above 97.5%, and the F-score is more than 97%.

Moreover, we evaluate the FPRs of recognizing specific type of vehicle dynamics. Figure 2.10b shows the box-plot of the FPRs for each type of vehicle dynamics. We can observe from Fig. 2.10b that the highest FPR is no more than 5% and the average FPR is as low as 2% over the three vehicle dynamics for all drivers. It shows that the proposed approaches can accurately sense vehicle dynamics in real driving environments.

2.4.4 Performance of Sensing Lane-Change

For evaluating the performance of sensing lane-change, we further divide the collected lane-change dataset into six specific categories, including two single lane-changes: towards left and right, and four sequential lane-changes: left-left, left-right, right-left and right-right.

Figure 2.11a shows the result of lane-change detection over six categories and the overall performance. In total, the accuracy of lane-change detection is 94.37%, while the overall precision, recall and F-score are 93.76%, 95.28% and 94.41%,

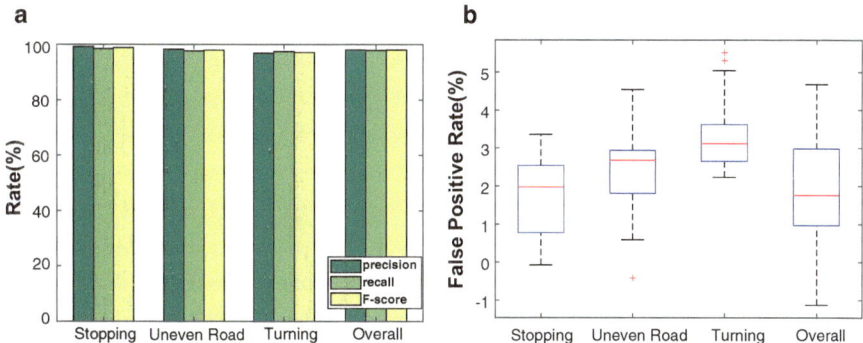

Fig. 2.10 The evaluation results on sensing different vehicle dynamics. (**a**) The precision, recall and F-score on sensing different vehicle dynamics. (**b**) Box plot of False Positive Rate on sensing different vehicle dynamics

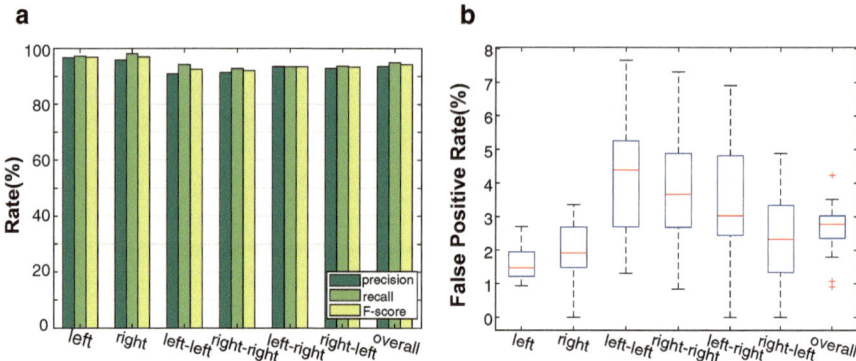

Fig. 2.11 The evaluation results on sensing lane-change for vehicles. (**a**) The precision, recall and F-score on sensing lane-change. (**b**) Box plot of False Positive Rate on sensing lane-change

respectively. For each of the six categories of lane-changes, the precision is no less than 90%, while the recall is above 92%, and the F-score is more than 91%. Moreover, for single lane-changes, the performance are slightly better than sequential lane-changes, as single lane-changes have more robust pattern in sensor readings. Since the single lane-change has the highest proportion in real driving environments, its highest accuracy contributes the most to the total performance of lane-change detection.

We further evaluate the FPRs of sensing specific categories of lane-change detection and the overall FPR. Figure 2.11b shows the box-plot of the FPRs for each category of lane-changes. It can be seen from Fig. 2.11b that the highest FPR is no more than 8% and the average FPR is as low as 3% over the six categories of lane-changes for all drivers. Similar to Fig. 2.11a, the FPR of single lane-changes are slightly lower than sequential lane-changes, as single lane-changes are more robust in detection. It shows that the proposed approaches in Sect. 2.3.4 can accurately sense different categories of lane-changes.

2.4.5 Performance of Sensing Instance Speed

For evaluating speed estimation accuracy of proposed approach in Sect. 2.3.5, we experiment the online approach with two types of roads: local road and elevated road, and further divide the collected data into different areas and different periods of day. We compare the estimated speed of the proposed approach with that of ground truth obtained from OBD-II adapter, and the result of GPS. Figure 2.12 presents the average estimation error.

It can be seen from Fig. 2.12 that the speed estimation leveraging all the reference points (i.e., All) has low errors and achieves better accuracy than that of GPS under

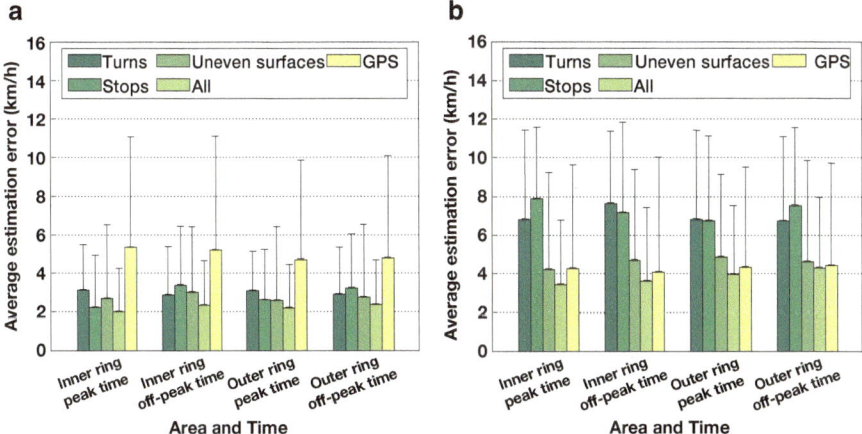

Fig. 2.12 The average estimation error of the vehicle speed. (**a**) Estimation error on local road. (**b**) Estimation error on elevated road

all types of roads and different periods of day. In particular, the average error of speed estimation is about 2.0 km/h on local roads, but it is up to 4.0 km/h on the elevated road. This is because there are less reference points on elevated roads than local roads.

Further, we evaluate the estimation accuracy of our system by using only one type of reference points. We find that the average estimation error on local road is still lower than of GPS even if only one type of reference points is used. However, the speed estimation using turns or stops is worse than that of GPS under elevated road and highways due to the fact that there are less turns and stops can be used as reference points. Still, we find that by using uneven road surfaces only, we can achieve comparable or better accuracy when comparing with GPS under all types driving roads. The result shows that our approach is accurate and robust enough in real driving environments.

2.5 Conclusion

In this chapter, we address the problem of enhancing off-the-shelf smartphone for sensing vehicle dynamics to support vehicular applications. We mainly employ two smartphone sensors, i.e., accelerometer and gyroscope, to achieve the goal. In particular, through digging into sensor readings, we present methods for sensing five different types of vehicle dynamics, which are the moving and stopping, driving on uneven road, turning, changing lanes and the instance speed of vehicles when driving. Our extensive experiments show that our approaches are validate in real driving environments.

Chapter 3
Sensing Vehicle Dynamics for Abnormal Driving Detection

3.1 Introduction

According to the statistics from World Health Organization (WHO), traffic accidents have become one of the top ten leading causes of death in the world [14]. Specifically, traffic accidents claimed nearly 3500 lives each day in 2014. Studies show that most traffic accidents are caused by human factors, e.g. drivers' abnormal driving behaviors [15]. Therefore, it is necessary to detect drivers' abnormal driving behaviors to alert the drivers or report Transportation Bureau to record them.

Although there has been works [16, 17] on abnormal driving behaviors detection, the focus is on detecting driver's status based on pre-deployed infrastructure, such as alcohol sensor, infrared sensor and cameras, which incur high installation cost. Since smartphones have received increasing popularities over the recent years and blended into our daily lives, more and more smartphone-based vehicular applications [3, 12, 18] are developed in Intelligent Transportation System. Driving behavior analysis is also a popular direction of smartphone-based vehicular applications. However, existing works [19, 20] on driving behaviors detection using smartphones can only provide a coarse-grained result using thresholds, i.e. distinguishing abnormal driving behaviors from normal ones. Since thresholds may be affected by car type and sensors' sensitivity, they cannot accurately distinguish the differences in various driving behavioral patterns. Therefore, those solutions cannot provide fine-grained identification, i.e. identifying specific types of driving behaviors.

Moving along this direction, we need to consider a fine-grained abnormal driving behaviors monitoring approach using smartphone sensors without requiring any additional hardwares. The fine-grained abnormal driving behaviors monitoring is able to improve drivers' awareness of their driving habits as most of the drivers are over-confident and not aware of their reckless driving habits. Additionally, some

© The Author(s), under exclusive licence to Springer International Publishing AG, 25
part of Springer Nature 2018
J. Yu et al., *Sensing Vehicle Conditions for Detecting Driving Behaviors*,
SpringerBriefs in Electrical and Computer Engineering,
https://doi.org/10.1007/978-3-319-89770-7_3

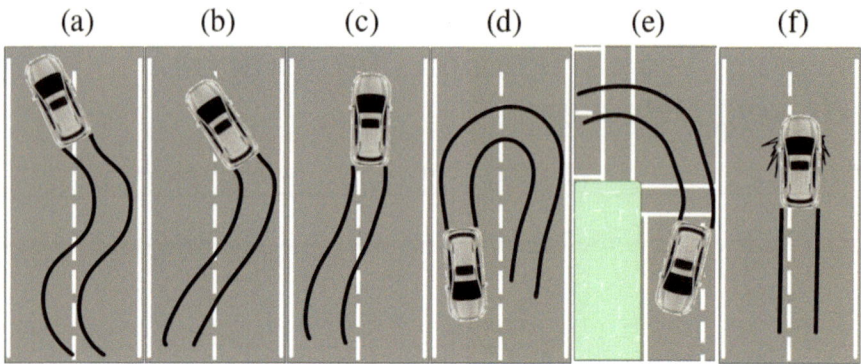

Fig. 3.1 Six types of abnormal driving behaviors: (**a**) weaving, (**b**) swerving, (**c**) sideslipping, (**d**) fast U-turn, (**e**) turning with a wide radius, (**f**) sudden braking

abnormal driving behaviors are unapparent and easy to be ignored by drivers. If we can identify drivers' abnormal driving behaviors automatically, the drivers can be aware of their bad driving habits, so that they can correct them, helping to prevent potential car accidents. Furthermore, if the results of the monitoring could be passed back to a central server, they could be used by the police to detect drunken-driving automatically or Vehicle Insurance Company to analyze the policyholders' driving habits.

According to [21], there are six types of abnormal driving behaviors defined, and they are illustrated in Fig. 3.1. *Weaving* (Fig. 3.1a) is driving alternately toward one side of the lane and then the other, i.e. serpentine driving or driving in S-shape; *Swerving* (Fig. 3.1b) is making an abrupt redirection when driving along a generally straight course; *Sideslipping* (Fig. 3.1c) is when driving in a generally straight line, but deviating from the normal driving direction; *Fast U-turn* (Fig. 3.1d) is a fast turning in U-shape, i.e. turning round (180°) quickly and then driving along the opposite direction; *Turning with a wide radius* (Fig. 3.2e) is turning cross an intersection at such an extremely high speed that the car would drive along a curve with a big radius, and the vehicle sometimes appears to drift outside of the lane, or into another line; *Sudden braking* (Fig. 3.2f) is when the driver slams on the brake and the vehicle's speed falls down sharply in a very short period of time.

In this chapter, we first set out to investigate effective features from smartphone sensors' readings that are able to depict each type of abnormal driving behavior. Through empirical studies of the 6-month driving traces collected from smartphone sensors of 20 drivers in a real driving environment, we find that each type of abnormal driving behaviors has its unique patterns on readings from accelerometers and orientation sensors. Effective features thus can be extracted to capture the patterns of abnormal driving behaviors. Then, we train those features through a machine learning method, *Support Vector Machine* (SVM), to generate a classifier model which could clearly identify each of driving behaviors. Based on the classifier model, we propose an abnormal *Driving behaviors Detection and iDentification* system, D^3 [22, 23], which can realize a fine-grained abnormal driving behaviors

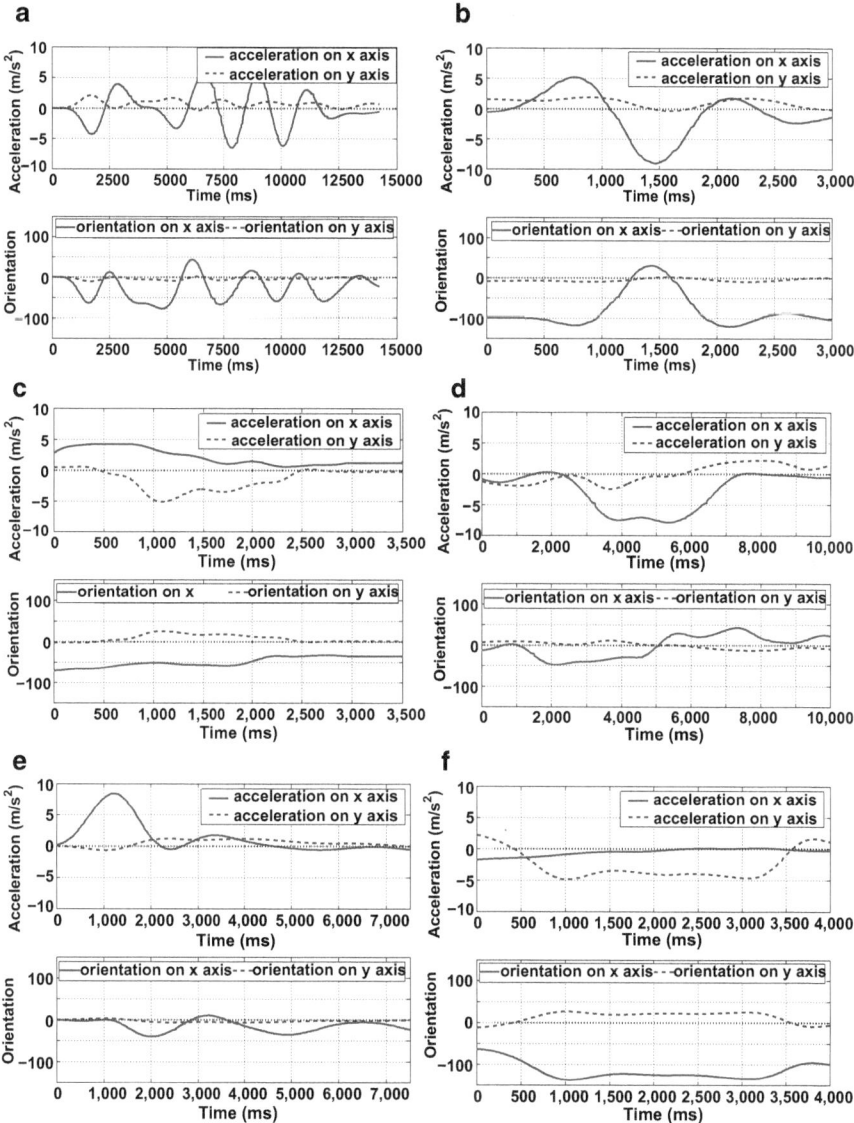

Fig. 3.2 The acceleration and orientation patterns of the six types of abnormal driving behaviors from an accelerometer and an orientation sensor's readings. (**a**) Weaving. (**b**) Swerving. (**c**) Sideslipping. (**d**) Fast U-turn. (**e**) Turning with a wide radius. (**f**) Sudden braking

monitoring in real-time using smartphone sensors. Our prototype implementation of D^3 on Android-based mobile devices verifies the feasibility of using D^3 in real driving environments.

The rest of the chapter is organized as follows. In Sect. 3.2, we analyze the acceleration and orientation patterns of the six specific types of abnormal driving behaviors. We present the design details of D^3 in Sect. 3.3. We evaluate the performance of D^3 and present the results in Sect. 3.4. Finally, we give the conclusion remarks in Sect. 3.5.

3.2 Driving Behavior Characterization

In this section, we first describe the data collection process for driving behavior samples from real driving environments. Then we analyze patterns of each type of driving behavior from smartphone sensors' readings.

3.2.1 Collecting Data from Smartphone Sensors

We develop an Android-based App to collect readings from the 3-axis accelerometer and the 3-axis orientation sensor. We align the two coordinate systems in the smartphone and in the vehicle by making the accelerometer's y-axis along the moving direction of the vehicle. Therefore, we could monitor the vehicle's acceleration and orientation by retrieving readings from the smartphone's accelerometer and orientation sensor.

We collect traces from the accelerometers and orientation sensors' readings on 20 drivers with distinct vehicles from January 11 to July 12, 2014. Each driver fixes a smartphone along with a Car Digital Video Recorder (DVR) in his/her vehicle within daily natural driving. The smartphone and Car DVR record the sensors' readings and all objective driving behaviors, respectively. The 20 drivers keep collecting data in their daily driving, including commute to work, shopping, touring and so on. Those 20 drivers live in different communities and they have different commute routes. On average, each driver may drive 60–80 km per day. 20 smartphones of five different types are used in our data collection, i.e. Huawei Honor3C, ZTE U809, SAMSUNG Nexus3, SAMSUNG Nexus4 and HTC sprint, four devices for each type. After that, we ask nine experienced drivers to watch the videos recorded by the Car DVR and recognize all types of abnormal driving behaviors from the 6-month traces, i.e. *Weaving*, *Swerving*, *Sideslipping*, *Fast U-turn*, *Turning with a wide radius* or *Sudden braking*. In total, we obtain 4029 samples of abnormal driving behaviors from the collected traces, which is viewed as the ground truth.

3.2.2 Analyzing Patterns of Abnormal Driving Behaviors

After high frequency noises are removed in the collected data using the low-pass filter, we can analyze the acceleration and orientation patterns of each type of abnormal driving behaviors. Let acc_x and acc_y be the acceleration on x-axis and y-axis, respectively. Let ori_x and ori_y be the orientation on x-axis and y-axis, respectively.

Weaving Figure 3.2a shows the acceleration and orientation patterns of weaving from an accelerometer and orientation sensor's readings. We observe from this figure that there is a drastic fluctuation on acc_x and this fluctuation continues for a period of time, while acc_y keeps smooth. Thus, both the standard deviation and the range of acc_x are very large and the time duration is long. The mean value of acc_x is around zero. In addition, the orientation values have similar patterns as acceleration values.

Swerving Figure 3.2b shows the acceleration and orientation patterns of swerving. Since swerving is an abrupt, instant behavior, the time duration is very short. When swerving occurs, there is a great peak on both acc_x and ori_x. Thus, the range and standard deviation of both acc_x and ori_x are large, and the mean value is not near zero. In addition, both acc_y and ori_y are flat during swerving.

Sideslipping Figure 3.2c shows the acceleration and orientation patterns of sideslipping. When sideslipping occurs, acc_y falls down sharply. Thus, the minimum value and mean value of acc_y are negative, and the range of acc_y is large. In addition, acc_x in sideslipping is not near zero. If the vehicle slips toward the right side, acc_x would be around a positive value, while if left, then negative. The mean value of acc_x thus is not near zero. When it comes to orientation, there are no obvious changes. Moreover, since sideslipping is an abrupt driving behavior, the time duration is short.

Fast U-Turn Figure 3.2d shows the acceleration and orientation patterns of fast U-turn. When a driver turns right or left fast in U-shape, acc_x rises quickly to a very high value or drops fast to a very low value, respectively. Moreover, the value would last for a period of time. The standard deviation of acc_x thus is large on the beginning and ending of a fast U-turn, the mean value of acc_x is far from zero and the range of acc_x is large. When it comes to acc_y, there are no obvious changes. Moreover, ori_x would pass over the zero point. Specifically, ori_x would change either from positive to negative or from negative to positive, depending on the original driving direction. Thus, the standard deviation and value range of ori_x would be large. The mean values in first half and second half of ori_x would be of opposite sign, i.e. one positive and the other negative. It may take a period of time to finish a fast U-turn, so its time duration is long.

Turning with a Wide Radius The acceleration and orientation patterns of turning with a wide radius are shown in Fig. 3.2e. When turning at an extremely high speed,

acc_x sees a high magnitude for a period of time, while the acc_y is around zero. Thus, the mean value of acc_x is far from zero and the standard deviation of acc_x is large. When it comes to orientation, ori_x sees a fluctuation, while ori_y keeps smooth. The standard deviation of ori_x thus is relatively large, and the mean value of ori_x is not near zero since the driving direction is changed. It may take a period of time to finish a turning with a wide radius, so the time duration is long.

Sudden Braking Figure 3.2f shows the acceleration and orientation patterns of sudden braking. When a vehicle brakes suddenly, acc_x remains flat while acc_y sharply downs and keeps negative for some time. Thus, the standard deviation and value range of acc_x are small. On acc_y, the standard deviation is large at the beginning and ending of a sudden braking and the range of acc_y is large. Moreover, there are no obvious changes on both ori_x and ori_y. Since sudden braking is an abrupt driving behavior, the time duration is short.

Normal Driving Behavior Normal driving behavior means smooth and safe driving with few and small fluctuations. Since there are few drastic actions in a normal driving behavior, the values on both acc_x and acc_y are not very large. So the mean, standard deviation, maximum and minimum values in acceleration on x-/y-axis are near zero. When it comes to orientation, a normal driving behavior presents smooth most of time. So the standard deviation and range of orientation are small.

Based on the analysis above, we find that each driving behavior has its unique features, e.g. standard deviation, mean, maximum, minimum, value range on acc_x, acc_y, ori_x and ori_y, as well as the time duration. Therefore, we could use those features to identify specific types of abnormal driving behaviors using machine learning techniques.

3.3   System Design

In this section, we present the design of our proposed system, D^3, which detects abnormal driving behaviors from normal ones and identifies different abnormal types using smartphone sensors. D^3 does not depend on any pre-deployed infrastructures and additional hardwares.

3.3.1   Overview

In our system, D^3, abnormal driving behaviors could be detected and identified by smartphones according to readings from accelerometers and orientation sensors. Figure 3.3 shows the architecture of D^3. The whole system is separated into offline part—*Modeling Driving Behaviors* and online part—*Monitoring Driving Behaviors*.

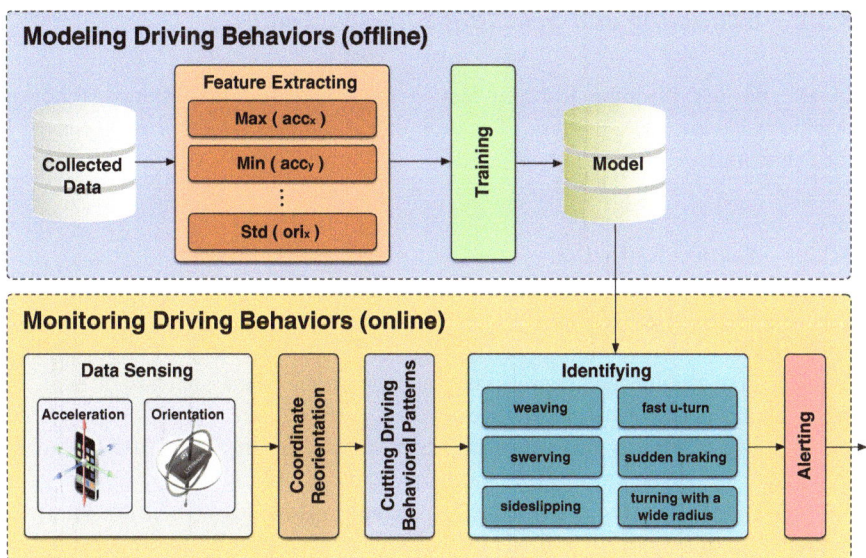

Fig. 3.3 System architecture

In the offline part, *Modeling Driving Behaviors*, D^3 trains a classifier model using machine learning techniques based on the collected data, which could identify specific types of driving behaviors. In the *Feature Extracting*, effective features are extracted from specific types of driving behavioral patterns on acceleration and orientation. Afterwards, the features are trained in the *Training* and a classifier model would be generated which can realize fine-grained identification. Finally, the classifier model is output and stored to *Model Database*.

The online part, *Monitoring Driving Behaviors*, is installed on smartphones which senses real-time vehicular dynamics to detect and identify abnormal driving behaviors. D^3 first senses the vehicles' acceleration and orientation by smartphone sensors. After getting real-time readings from the accelerometer and the orientation sensor, the *Coordinate Reorientation* is performed to align the smartphone's coordinate system with the vehicle's using the method in Sect. 2.2.1. Then, in the *Cutting Driving Behavioral Patterns*, the beginning and ending of a driving behavior are found out from accelerometer and orientation sensor's readings. Afterwards, in *Identifying*, D^3 extracts features from patterns of the driving behaviors, then identifies whether one of the abnormal driving behaviors occurs based on the classifier model trained in *Modeling Driving Behaviors*. Finally, if any of the abnormal driving behaviors were identified, a warning message would be sent to receivers by the *Alerting*.

3.3.2 Extracting and Selecting Effective Features

In D^3, we use machine learning techniques to identify fine-grained abnormal driving behaviors. The process of feature extraction and selection is discussed in the following.

3.3.2.1 Feature Extraction

When machine learning algorithms are processed, representative tuple of features rather than raw data is a more effective input. Thus, it is necessary to extract effective features from driving behavioral patterns. According to the analysis in Sect. 3.2, each driving behavior has its unique patterns on acc_x, acc_y, ori_x, ori_y and time duration (t). The main difference between various driving behaviors lies in the maximum, minimum, value range, mean, and standard deviation of acc_x, acc_y, ori_x and ori_y and t. Therefore, those values can be used as features for training. However, not all of them are equally effective for abnormal driving behaviors' detection and identification.

3.3.2.2 Feature Selection

In order to select the really effective features, we analyze the collected traces. Figure 3.4 shows some of the effective features which distinguish abnormal driving behaviors from normal ones and distinguish weaving from the other five abnormal driving behaviors.

Figure 3.4a shows the difference between normal and abnormal driving behaviors in a two-dimensional features tuple (i.e. $range_{acc,x}$ and $range_{acc,y}$). It can be seen that the two features can clearly discriminate normal and abnormal driving behaviors. Therefore, we manage to distinguish abnormal driving behaviors from normal ones with only two features.

In fact, additionally to the two features shown in Fig. 3.4a, some other combinations of a two-dimensional features tuple (i.e. any two out of t, $max_{ori,x}$, $max_{ori,y}$, $\sigma_{ori,x}$, $\sigma_{ori,y}$, $\sigma_{acc,x}$, $range_{acc,x}$, $min_{acc,y}$ and $range_{acc,y}$) also manage to distinguish abnormal driving behaviors from normal ones.

Although we can distinguish abnormal driving behaviors from normal ones using a two-dimensional features tuple, we fail to differentiate the six types of abnormal behaviors from each other only using two-dimensional features. As the example shown in Fig. 3.4a, the six types of abnormal driving behaviors are mixed with each other. Nevertheless, they could be differentiated pairwise with a two-dimensional features tuple. In other words, although the six abnormal driving behaviors cannot be differentiated from each other at the same time, any two among them can be differentiated intuitively by a two-dimensional features tuple. Taking weaving for example (see Fig. 3.4b–f), weaving can be distinguished from the other five

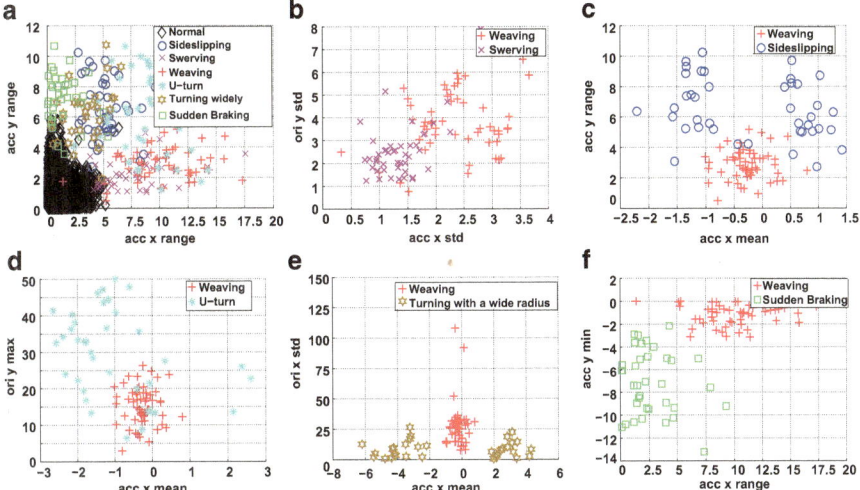

Fig. 3.4 Some effective features for identifying normal driving behavior from abnormal ones and weaving behavior from other five abnormal driving behaviors. (**a**) Normal vs. abnormal. (**b**) Weaving vs. swerving. (**c**) Weaving vs. sideslipping. (**d**) Weaving vs. fast U-turn. (**e**) Weaving vs. turning with a wide radius. (**f**) Weaving vs. sudden braking

abnormal driving behaviors using a two-dimensional features tuple. For instance, in Fig. 3.4b, weaving and swerving can be discriminated from each other using $\sigma_{ori,y}$ and $\sigma_{acc,x}$. Similarly, other abnormal driving behaviors can also be pairwise discriminated using two-dimensional features tuples.

Based on the collected traces, we investigate all possible pairwise cases. In each case, we find out several effective features conductive to distinguishing one driving behavior from another. Finally, we identify 16 effective features that are able to capture the patterns of different types of abnormal driving behaviors, as listed in Table 3.1.

3.3.3 Training a Fine-Grained Classifier Model to Identify Abnormal Driving Behaviors

After feature extracting, we obtain a tuple of features for each driving behavior. Then a classifier model is trained based on the tuples for all driving behaviors through machine learning techniques [24] to identify various driving behaviors. We use the multi-class SVM [25, 26] to train the classifier model. For each driving behavior, the input into SVM is in the form of <16-dimensional features, label>, where the 16-dimensional features are the tuples obtained from the *Feature Extracting* and the label is the type of the driving behavior.

Table 3.1 Features extracted

Feature	Description
$range_{acc,x}$	Subtraction of maximum minus minimum value of acc_x
$range_{acc,y}$	Subtraction of maximum minus minimum value of acc_y
$\sigma_{acc,x}$	Standard deviation of acc_x
$\sigma_{acc,y}$	Standard deviation of acc_y
$\sigma_{ori,x}$	Standard deviation of ori_x
$\sigma_{ori,y}$	Standard deviation of ori_y
$\mu_{acc,x}$	Mean value of acc_x
$\mu_{acc,y}$	Mean value of acc_y
$\mu_{ori,x}$	Mean value of ori_x
$\mu_{ori,y}$	Mean value of ori_y
$\mu_{acc,x,1}$	Mean value of 1st half of acc_x
$\mu_{acc,x,2}$	Mean value of 2nd half of acc_x
$max_{ori,x}$	Maximum value of ori_x
$max_{ori,y}$	Maximum value of ori_y
$min_{acc,y}$	Minimum value of acc_y
t	Time duration between the beginning and the ending of a driving behavior

The cores in SVM are the *kernel* and the *similarity function*. A *kernel* is a landmarks, and the *similarity function* computes the similarity between an input example and the kernels. Specifically, assume that our training set contains m samples, and each sample is 16-dimensional (i.e. the 16-dimensional features), denoted by

$$x^{(i)} = (x_1^{(i)}, x_2^{(i)}, \cdots, x_{16}^{(i)}), \qquad i = 1, 2, \cdots, m, \qquad (3.1)$$

where $x^{(i)}$ is the ith sample, and $x_j^{(i)}$ means the jth feature of $x^{(i)}$. When SVM starts, all input samples $(x^{(1)}, x^{(2)}, \cdots, x^{(m)})$ are selected as kernels, recorded as $l^{(1)}, l^{(2)}, \cdots, l^{(m)}$. Note that $x^{(i)} = l^{(i)}$ for $i = 1, 2, \cdots, m$. Afterwards, for each sample, SVM compute its similarity between the kernels by

$$f_j^{(i)} = e^{-\frac{||x^{(i)} - l^{(j)}||^2}{2\sigma^2}}, \qquad i, j = 1, 2, \cdots, m, \qquad (3.2)$$

where $f_j^{(i)}$ is the similarity between input sample $x^{(i)}$ and the kernel $l^{(j)}$, σ is a parameter defined manually, and $||x^{(i)} - l^{(j)}||^2$ is the distance between $x^{(i)}$ and $l^{(j)}$ calculated by

$$||x^{(i)} - l^{(j)}||^2 = \sum_{k=1}^{16}(x_k^{(i)} - l_k^{(j)})^2,$$

$$i, j = 1, 2, \cdots, m.$$

(3.3)

In SVM, those m 16-dimensional input samples (i.e. $x^{(1)}, x^{(2)}, \cdots, x^{(m)}$) would be converted into m m-dimensional similarity features (i.e. $f^{(1)}, f^{(2)}, \cdots, f^{(m)}$), since for each $x^{(i)}$, the similarity between $x^{(i)}$ and any $l^{(j)}$ in $l^{(1)}, l^{(2)}, \cdots, l^{(m)}$ are calculated by Eq. (3.2). With the new features $f = (f^{(1)}, f^{(2)}, \cdots, f^{(m)})$, a cost function $J(\theta)$ (see Eq. (3.4)) calculated from f would be minimized to find optimal θ.

$$J(\theta) = C \sum_{i=1}^{m} y^{(i)} cost_1(\theta^T f^{(i)}) + (1 - y^{(i)})cost_0(\theta^T f^{(i)})$$
$$+ \frac{1}{2} \sum_{j=1}^{m} \theta_j^2,$$

(3.4)

where C is a parameter defined manually, $y^{(i)}$ is the label of ith input example (i.e. the label of $x^{(i)}$), θ^T means θ transpose and

$$cost_1(\theta^T f^{(i)}) = \log(\frac{1}{1 + e^{-\theta^T f^{(i)}}}),$$

$$cost_0(\theta^T f^{(i)}) = \log(1 - \frac{1}{1 + e^{-\theta^T f^{(i)}}})$$

(3.5)

and

$$\theta^T f^{(i)} = \theta_1 f_1^{(i)} + \theta_2 f_2^{(i)} + \cdots + \theta_m f_m^{(i)}.$$

(3.6)

The classifier model would be finally determined by the optimal θ. In a word, SVM trains the inputs and then output a classifier model which conducts fine-grained identification to the six types of abnormal driving behaviors.

3.3.4 Detecting and Identifying Abnormal Driving Behaviors

After we obtain a classifier model, we are able to detect and identify abnormal driving behaviors in real driving environments using the model. In order to identify current driving behavior using the model, we should input features extracted from patterns of a driving behavior. D^3 thus need to determine the beginning and ending of the driving behavior first, i.e., cutting patterns of the driving behavior. Figure 3.5

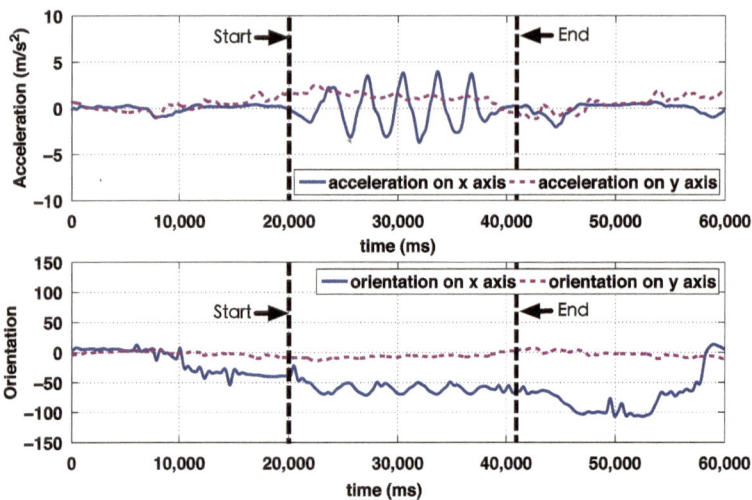

Fig. 3.5 The acceleration and orientation patterns of 1 min driving behaviors

shows the readings from a smartphone' accelerometer and orientation sensor on x-axis and y-axis in a 1 min driving, which contains a weaving behavior. In Fig. 3.5, the weaving behavior is sensed from its beginning to ending.

The method of sensing the beginning and ending of a driving behavior is proposed based on an analysis on the acceleration and orientation patterns of all types of driving behaviors. Specifically, when an abnormal driving behavior begins, the standard deviation of either the acceleration or the orientation values sharply rise to and keep a relatively high value until the ending, while in most normal driving behaviors, the standard deviation always presents as low and smooth. Moreover, during an abnormal driving behavior, the magnitude of acceleration on either x-axis or y-axis presents an extremely high value, as illustrated in Sect. 3.2. But when driving normally, the magnitude of accelerations seldomly reaches to such a high value.

Therefore, it is simple but effective that we monitor the standard deviation of acceleration and orientation as well as the magnitude of acceleration of the vehicle to cut patterns of driving behaviors. In real driving environments, we retrieve readings from smartphones' accelerometers and orientation sensors and then compute their standard deviation as well as mean value in a small window. Under normal driving, D^3 compares the standard deviation and the mean value with some thresholds to determine whether an abnormal driving behavior begins. The window size and thresholds can be learned from the collected data. After the beginning of a driving behavior is found out, D^3 continues to check the standard deviation and mean value to determine whether the driving behavior ends.

After cutting patterns of a driving behavior, effective features can be extracted from the driving behavioral patterns and then sent to the classifier model. Finally, the model outputs a fine-grained identification result. If the result denotes the normal driving behavior, it is ignored, and if it denotes any one of abnormal driving behaviors, D^3 sends a warning message.

3.4 Evaluations

In this section, we first present the prototype of D^3, then evaluate the performance of D^3 in real driving environments.

3.4.1 Setup

We implement D^3 as an Android App and install it on smartphones (listed in Sect. 3.2.1). Figure 3.6 shows the user interface of D^3 and testbeds in vehicles. D^3 is running by 20 drivers with distinct vehicles in real driving environments to collect the data for evaluation. The 20 drivers driver car within daily natural driving and are not aware of the smartphones are embedded with D^3. Meanwhile, Car DVRs are used to record driving behaviors and nine experienced drivers are asked to recognize abnormal driving behaviors as ground truth. After a 4-month data collection (i.e. July 21 to November 30, 2014, using the same method of collecting data as described in Sect. 3.2.1), we obtain a test set with 3141 abnormal driving behaviors to evaluate the performance of D^3.

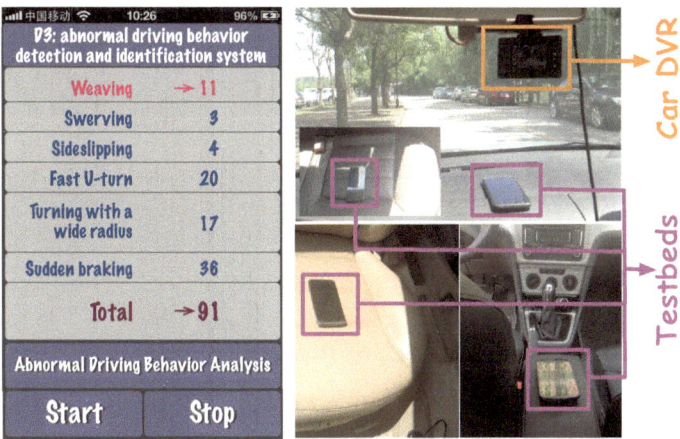

Fig. 3.6 User interface of D^3 and testbeds

3.4.2 Metrics

To evaluate the performance of D^3, we define the following metrics based on the True Positive (TP), True Negative (TN), False Positive (FP) and False Negative (FN).

- **Accuracy**: The probability that the identification of a behavior is the same as the ground truth.
- **Precision**: The probability that the identifications for behavior A is exactly A in ground truth.
- **Recall**: The probability that all behavior A in ground truth are identified as A.
- **False Positive Rate (FPR)**: the probability that a behavior of type Not A is identified as A.

In the following subsections, we investigate the impact of various factors to D^3 and present the details.

3.4.3 Overall Performance

The performance of D^3 is evaluated by three levels, i.e. total accuracy, detecting abnormal vs. normal driving behaviors and identifying fine-grained driving behaviors.

3.4.3.1 Total Accuracy

Total accuracy is the ratio of correct identifications to total identifications, containing identifications for the six types of abnormal driving behaviors as well as the normal. The total accuracy for each driver is evaluated and the result is shown in Fig. 3.7. It can be seen that all of the 20 drivers achieve high total accuracies. Among

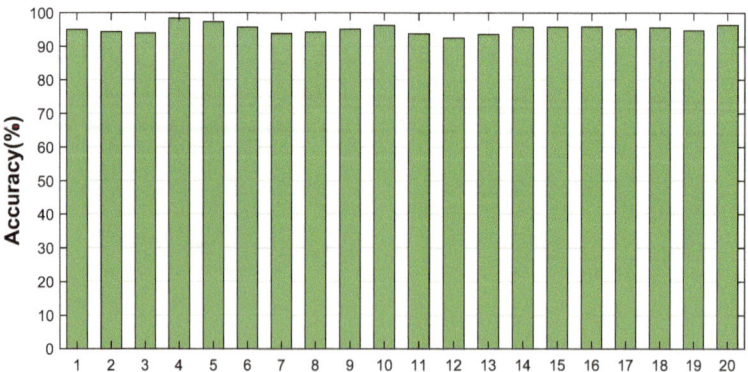

Fig. 3.7 The total accuracy of D^3 over 20 drivers

the 20 drivers, the lowest total accuracy is 92.44%. On average, D^3 achieves a total accuracy of 95.36%.

3.4.3.2 Detecting the Abnormal vs. the Normal

In this level, we treat all types of abnormal driving behaviors as one type (i.e. *Abnormal*), and merely identify whether a driving behavior is abnormal or normal. As is shown in Table 3.2, D^3 performs so excellent that almost all abnormal driving behaviors are identified, with only 6 out of 3141 omitted. In other words, D^3 could identify abnormal driving behaviors vs. normal ones with a recall of 99.84%. In addition, none of normal driving behaviors is identified as abnormal one, i.e. with 100% precision and 0 FPR.

3.4.3.3 Identifying Abnormal Driving Behaviors

D^3 also realizes fine-grained identification, i.e., discriminates *Weaving*, *Swerving*, *Sideslipping*, *Fast U-turn*, *Turning with a wide radius* and *Sudden braking*. Table 3.2 shows the identification results. The accuracy for identifying each of the six abnormal driving behaviors is no less than 94%, the precision is above 85%, and the recall is more than 70%. The FPRs for identifying all types of abnormal driving behaviors are no more than 2%. The results show that D^3 is an high-accurate system to identify various abnormal driving behaviors.

Moreover, we evaluate FPRs of identifying specific abnormal types. Figure 3.8 shows a box-plot of the FPRs for each type of abnormal driving behaviors and the overall FPR. As is shown in the figure, the highest FPR of identifying specific abnormal type is less than 2.5% and the overall FPR is around 0.9%, which shows that D^3 could implement fine-grained identification with few false alarms. In addition, D^3 performs better when identifying weaving, sideslipping, turning with a wide radius and fast U-turn than identifying swerving and sudden braking. This

Table 3.2 Accuracy evaluation

Behavior	Accuracy (%)	Precision (%)	Recall (%)	FPR (%)
Normal	99.84	98.80	100.00	0.19
Abnormal	94.81	100.00	99.80	0.00
Weaving	98.43	92.55	87.87	0.63
Swerving	97.94	92.29	94.15	1.39
Sideslipping	98.60	87.96	71.43	0.37
Fast U-turn	98.49	85.71	76.00	0.54
Turning with a wide radius	98.68	89.30	92.72	0.86
Sudden braking	95.74	97.88	99.04	1.93

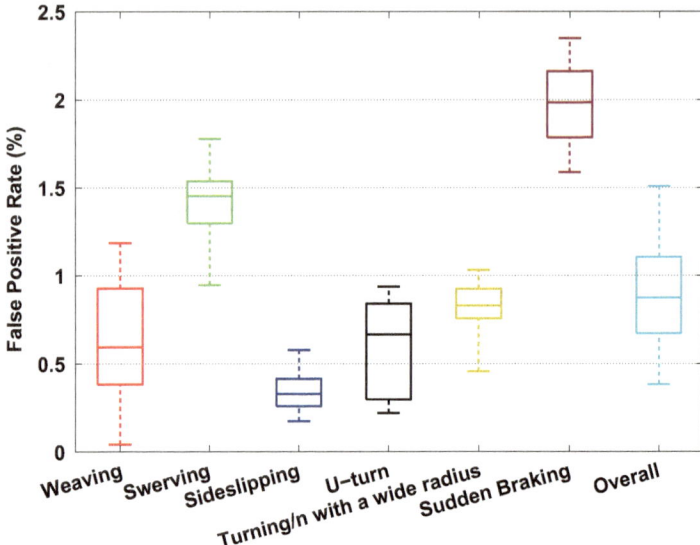

Fig. 3.8 Box plot of FPR of identifying specific types of driving behaviors

is because the patterns of the former ones are more distinct than that of the latter. However, the performance of identifying swerving, sideslipping and turning with a wide radius is more stable than identifying other abnormal driving behaviors since they have smaller standard deviations. This is because the patterns of the former ones are more stable than that of the latter.

3.4.4   Impact of Training Set Size

According to Sect. 3.2.1, we collect 4029 abnormal driving behaviors in total for training. The training set size (i.e. the number of training samples) may have an impact on the training results so that it may affect the performance of D^3. We thus evaluate the impact of the training set size. The results are shown in Fig. 3.9. From the figure, we observe that the more training samples there are, the better performance D^3 has. When we use 280 training samples for turning with a wide radius, sideslipping, 300 sudden braking samples, 350 swerving samples and 380 training samples for fast U-turn and weaving, respectively, D^3 could identify each specific type of driving behavior with an accuracy close to 100%. In order to guarantee the performance of D^3, we use as many training samples as possible.

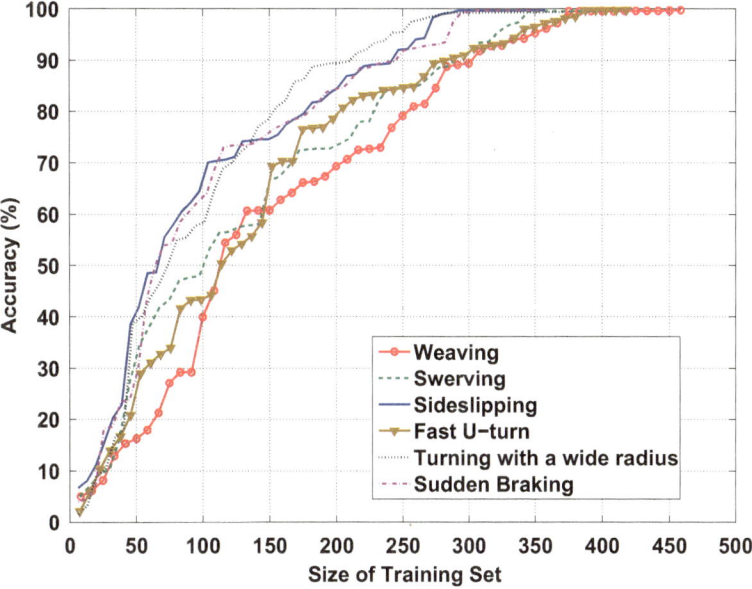

Fig. 3.9 Total accuracy under different sizes of training set

3.4.5 Impact of Traffic Conditions

The traffic conditions may affect the drivers' driving behaviors and further affect the performance of D^3. We analyze traces during peak time and off-peak time respectively to evaluate the impact of traffic conditions. Figure 3.10 shows the accuracies of identifying specific types of abnormal driving behaviors during peak and off-peak time. It can be seen that D^3 achieves good accuracy during both time periods, and the accuracy in off-peak time is slightly higher than that in peak time. This is because during peak time, the vehicles perform less drastic actions due to traffic jams. So some abnormal driving behaviors present restrained patterns during peak time. Different types of abnormal driving behaviors thus are much easier to be mistaken by each other and even be mistaken as normal driving behaviors. Nevertheless, during off-peak time, the patterns of all types of driving behaviors are performed more obvious. So different types of abnormal driving behaviors are more distinguishable.

3.4.6 Impact of Road Type

Drivers could perform abnormal driving behaviors on highway or local road, thus we further investigate the impact of the two road types on the performance of D^3.

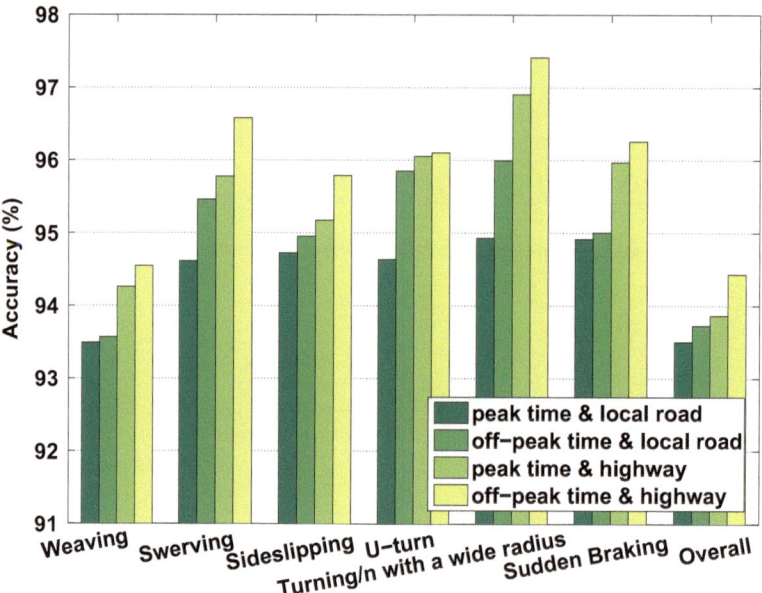

Fig. 3.10 Accuracy under different traffic conditions and rode types

Figure 3.10 shows how road types affect the accuracy of identifying various types of abnormal driving behaviors. It can be seen that D^3 achieves good accuracy both on highway and local road, but the accuracy is slightly higher on highway than that on local road. This is because the better road condition on highway could reduce the fluctuations caused by bumpy surfaces. Since highway is more smooth and has less slopes compared with local road, there are less disturbances then. In addition, there are less curves and no traffic light stops on the highway, so when driving normally on the highway, drivers have less chance to perform drastic actions. As a result, D^3 can achieve a better performance on highway than that on local road.

3.4.7   Impact of Smartphone Placement

Smartphones could be arbitrarily placed in vehicles, we thus investigate the impact of smartphone placement. In our experiments with 20 vehicles, smartphones are fixed on instrument panel, cupholder on the center console, front passenger seat, or left rear passenger seat, where smartphone sensors' y-axis is aligned along the moving direction of vehicles, or on arbitrary placement (i.e. smartphones are put in the driver's pocket and its pose could be arbitrary). Figure 3.11 shows the CDF of FPRs of fine-grained identifications under different smartphone placements. It can be seen that D^3 can achieve low FPRs under all smartphone placements, which

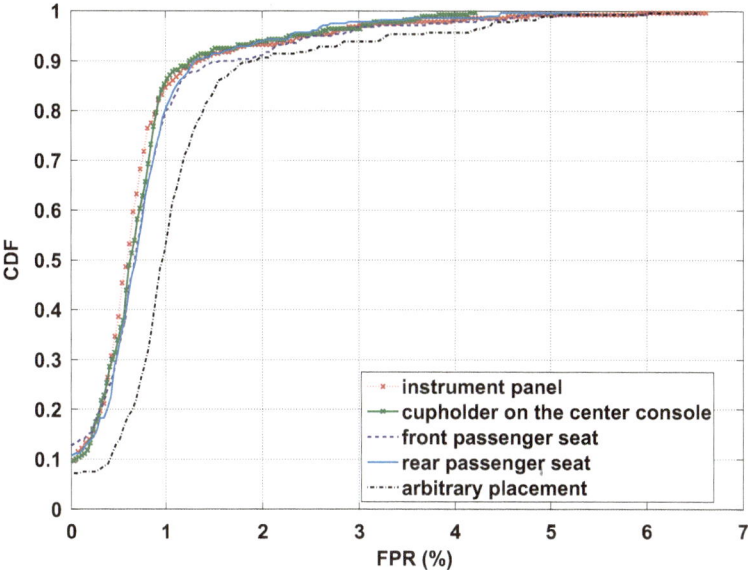

Fig. 3.11 CDF of FPR under different smartphone placements

shows D^3 performs excellent wherever the smartphone is placed in a vehicle. Although there is slightly higher FPR under arbitrary placement because of errors in the coordinate reorientation process, a FPR of less than 2% in 90% of the cases is still a good result.

3.5 Conclusion

In this chapter, we address the problem of performing abnormal driving behaviors detection (coarse-grained) and identification (fine-grained) to improve driving safety. In particular, we propose a system, D^3, to detect and identify specific types of abnormal driving behaviors by sensing the vehicle's acceleration and orientation using smartphone sensors. Compared with existing abnormal driving detection systems, D^3 not only implements coarse-grained detections but also conducts fine-grained identifications. To identify specific abnormal driving behaviors, D^3 trains a multi-class classifier model through SVM based on the acceleration and orientation patterns of specific types of driving behaviors. To obtain effective training inputs, we extract 16 effective features from driving behavioral patterns collected from the 6-month driving traces in real driving environments. The extensive experiments driving in real driving environments in another 4 months show that D^3 achieves high accuracy when detecting and identifying abnormal driving behaviors.

Chapter 4
Sensing Driver Behaviors for Early Recognition of Inattentive Driving

4.1 Introduction

Inattentive driving [27] is a significant factor in distracted driving and is associated with a large number of car accidents. According to statistics, in 2014, 3179 people were killed and 431,000 were injured in the United States alone in motor vehicle crashes involving inattentive drivers [28]. National Highway Traffic Safety Administration (NHTSA) is working to reduce the occurrence of inattentive driving and raise awareness of the dangers of inattentive driving [29]. However, recent research [30] shows that many inattentive driving events are unapparent and thus easy to be ignored by drivers. Most drivers fail to realize themselves as inattentive while driving. Therefore, it is desirable to build an inattentive driving recognition system to alert drivers in real time, helping to prevent potential car accidents and correct drivers' bad driving habits.

There have been existing studies on detecting abnormal driving behaviors [16, 17] including inattentive, drowsy and drunk drinking. These studies detect driver's status based on pre-deployed infrastructure, such as cameras, infrared sensors, and EEG devices, incurring high cost. In recent years, with the increasing popularity of smartphones, more and more smartphone-based applications [10, 22, 31] are developed to detect driving behaviors using sensors embedded in smartphones, such as accelerator, gyroscope, and camera. However, most of these investigations on driving behavior detection using smartphones can only provide a detection result after a specific driving behavior is finished, making it less helpful to alert drivers and avoid car accidents.

Among all kinds of dangerous driving behaviors, inattentive driving is the most common one but also easily to be ignored by drivers. Thus, early recognition of inattentive driving is the key to alert drivers and reduce the possibility of car accidents. Our objective is to build a system for early recognition of inattentive

J. Yu et al., *Sensing Vehicle Conditions for Detecting Driving Behaviors*, SpringerBriefs in Electrical and Computer Engineering, https://doi.org/10.1007/978-3-319-89770-7_4

driving using existing smartphone sensors. According to the judicial interpretation of inattentive driving [27], there are four most commonly occurring events of inattentive driving, i.e. *Fetching Forward*, *Picking up Drops*, *Turning Back* and *Eating & Drinking*. Our goal is to recognize these most common inattentive driving events and alert drivers as early as possible to prevent drivers from continuing these behaviors dangerous to driving safety. Our work is grounded on the basic physics phenomenon that human actions may lead to Doppler shifts of audio signals [32] to recognize different inattentive driving events. To realize the inattentive driving recognition leveraging audio signals, we face several challenges in practice. First, the unique pattern of each type of inattentive driving needs to be distinguished. Second, any inattentive driving event should be recognized as early as possible under the guarantee of a high recognition accuracy. Finally, the solution should be effective in real driving environments and computational feasible on smartphones.

In this chapter, we first investigate the patterns of Doppler shifts of audio signals caused by inattentive driving events. Through empirical studies of the driving traces collected from real driving environments, we find that each type of inattentive driving event exhibits an unique pattern on Doppler profiles of audio signals. Based on the observation, we propose an *E*arly *R*ecognition system, *ER* [33, 34], which aims to recognize inattentive driving events at an early stage and alert drivers in real time for safe driving. In ER, effective features of inattentive driving events on audio signals collected by smartphones are first extracted through *Principal Components Analysis (PCA)*. To improve the recognition accuracy of driving events, we train these features through a machine learning method to generate binary classifiers for every pair of inattentive driving events, and propose a *modified vote mechanism* to form a multi-classifier for all inattentive driving events based on the binary classifiers. In this work, the training is performed based on 3-month driving traces in real driving environments involving eight drivers. Furthermore, to detect the inattentive driving at an early stage, we first analyze the relationship between the completion degree and time duration for each type of inattentive driving event, and then exploit the relationships to turn the multi-classifier into a *Gradient Model Forest* for early recognition. Our extensive experiments validate the accuracy and the feasibility of our system in real driving environments.

The rest of the chapter is organized as follows. Patterns of inattentive driving events on Doppler profiles of audio signals are analyzed in Sect. 4.2. Section 4.3 presents the design details of ER system. We evaluate the performance of ER and present the results in Sect. 4.4. Finally, we give our solution remarks in Sect. 4.5.

4.2 Inattentive Driving Events Analysis

In this section, we first give a brief introduction to inattentive driving events, and then analyze patterns of these events on Doppler profiles of audio signals.

4.2.1 Defining Inattentive Driving Events

Drivers are encountered with a variety of road hazards because of their unawareness of being in negligent driving states, such as eating or picking up drops while driving. These inattentive driving events are potentially posing drivers in danger. According to the judicial interpretation [27], there are four types of the most commonly occurring inattentive driving events of drivers, as shown in Fig. 4.1.

Fetching Forward Drivers fetch out to search widgets like keys, car audio consoles, etc.

Eating or Drinking Drivers eat snacks or replenishing water when driving.

Turning Back Drivers intend to take care of their children in rear seat, or turn around searching for bags or packages.

Picking Up Drops Drivers are likely to pick up dropped keys or other objects when driving where their heads temporarily moves away from the front sight.

Through analyzing the above four inattentive driving events, we realize each driving event is not a transient action, but a consecutive action lasts for a time period. For example, Fig. 4.1a shows a Fetching Forward event, which can be demonstrated

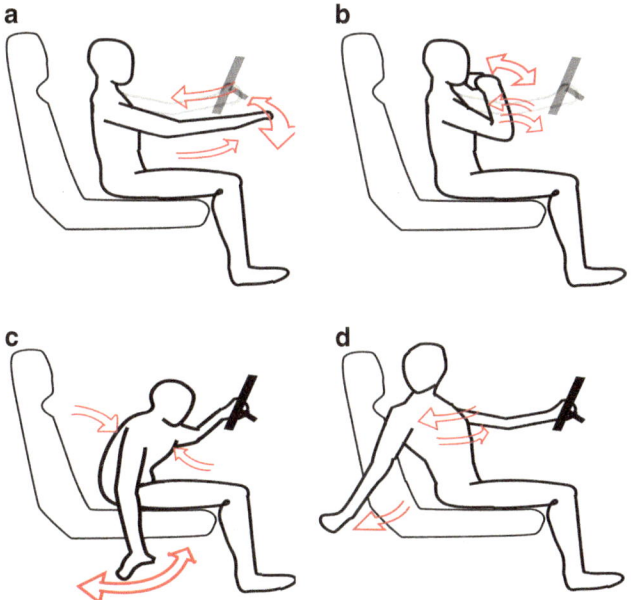

Fig. 4.1 Illustration of inattentive driving events. (**a**) Fetching Forward. (**b**) Eating & Drinking. (**c**) Picking up Drops. (**d**) Turning Back

as stretching out to reach the deck, searching, and stretching retrieved to normal condition. Our work is to detect these consecutive inattentive driving events in real time and try to recognize these events at the early stage, so as to alert drivers as early as possible.

4.2.2 Analyzing Patterns of Inattentive Driving Events

We utilize the Doppler shifts of audio signals to recognize inattentive driving events. *Doppler shift* (or Doppler effect) is the change in the observed frequency of a wave as the transmitter of the wave moves relative to the receiver. In our study, we utilize the speaker of a smartphone to transmit audio signals, then the signals are reflected by the objects they encounter in the environment and finally received by the microphone of the smartphone. Since the transmitter and receiver are in the same smartphone, the object that reflects the audio signals from the transmitter can be considered as a virtual transmitter that generates the reflected signals. Specifically, an object moving at speed v and angle θ to a smartphone brings a frequency change:

$$\Delta f = \frac{2v cos(\theta)}{c} \times f_0, \tag{4.1}$$

where c and f_0 denote the speed and frequency of the audio signals. Thus, we can leverage Doppler shift of the reflected audio signals to capture the movement pattern of the objects with respect to the smartphone.

We recruit five volunteers to perform four inattentive driving events depicted in Fig. 4.1 while driving in relatively safe area. The experiments are conducted by generating continuous pilot tones from speakers and then collect the audio signals from microphones on smartphones.

When selecting the frequency of audio signals to use, we take two factors into consideration, i.e. background noise and unobtrusiveness. According to [35], frequency range from 50 to 15,000 Hz covers almost all naturally occurring sounds, and human hearing becomes extremely insensitive to frequencies beyond 18 kHz. Thus, we could straightforwardly filter the background noise and eliminate the effects for people by locating our signal above 18 kHz. Furthermore, a higher frequency results in a more discernible Doppler shift confined by Eq. (4.1), and most smartphone speaker systems only can product audio signals at up to 20 kHz. Taking all above analysis into account, $f_0 = 20$ kHz is selected as our frequency of pilot tone, through which we sample raw data from given inattentive driving events at the rate of 44.1 kHz, which is the default sampling rate of audio signals under 20 kHz. Then we transform it into frequency domain using *2048-points Fast Fourier Transform (FFT)* for appropriate computation complexity and relative high frequency resolution.

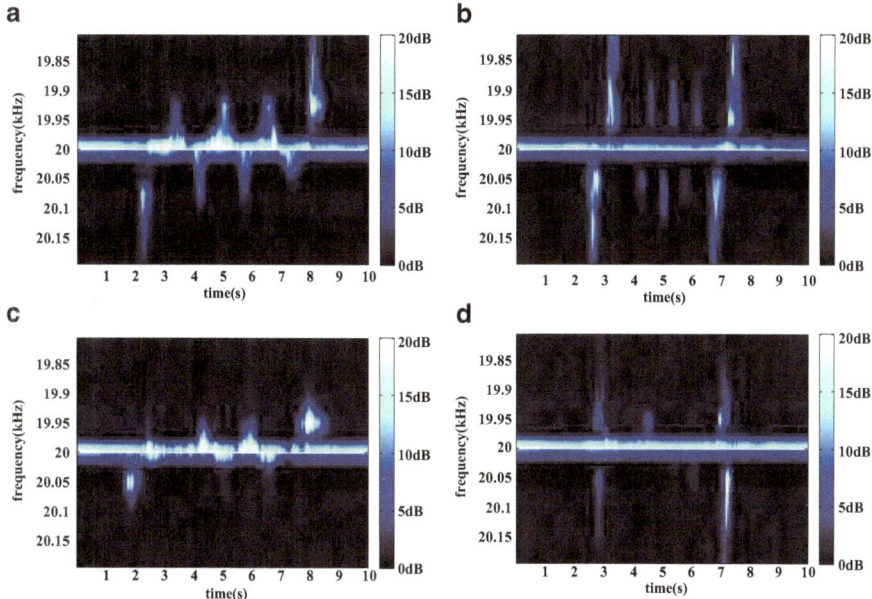

Fig. 4.2 Frequency-time Doppler profiles of four typical types of inattentive driving events. (**a**) Forward Fetching. (**b**) Eating & Drinking. (**c**) Picking up Drops. (**d**) Turning Back

Figure 4.2 shows the structure of Doppler profiles of the four typical types of inattentive driving events. From Fig. 4.2, it can be seen that although the four profiles share the similarity that they all consist of several positive and negative Doppler shifts, the patterns are different across the four events in frequency range, energy amplitude, etc.

From above analysis, we find that each type of inattentive driving events has unique patterns on the structure of Doppler profiles. Although some existing works, [22, 23, 32], present human action recognition methods based on the unique patterns of actions already, the recognition can only be done after actions finished, which is acceptable for transient actions like single gestures in [32], but not good enough for consecutive actions like inattentive driving events here because it is too late to alert drivers after the driving events finished in a driving security warning system. Our goal is to recognize inattentive driving events as early as possible and alert drivers timely.

4.3   System Design

In order to monitor inattentive driving events effectively and efficiently, we present an early recognition system, ER, which can alert drivers as early as possible when they are performing inattentive driving events. ER does not depend on any pre-deployed infrastructure and additional hardware.

4.3.1   System Overview

ER can recognize inattentive driving events through analyzing patterns of Doppler profiles of audio signals over time. The work flow of ER is shown in Fig. 4.3. The whole system is divided into offline part—*Modeling Inattentive Driving Events*, and online part—*Monitoring Inattentive Driving Events*.

In the offline part, for different types of inattentive driving events, effective features are extracted from the Doppler profiles of audio signals collected in real driving environments. Then, we train these features through machine learning methods to generate binary classifiers for every pair of inattentive driving events, and propose a *modified vote mechanism* to form a multi-classifier for all inattentive driving events along with other driving behaviors based on them. Afterwards, the multi-classifier model is turned into a *gradient model forest* for realizing early recognition, which is stored in the database.

In the online part, ER senses real-time audio signals generated by speakers and received by microphones. The audio signals are first transformed through FFT to Doppler profiles. Then, ER detects the beginning of an event and continuously

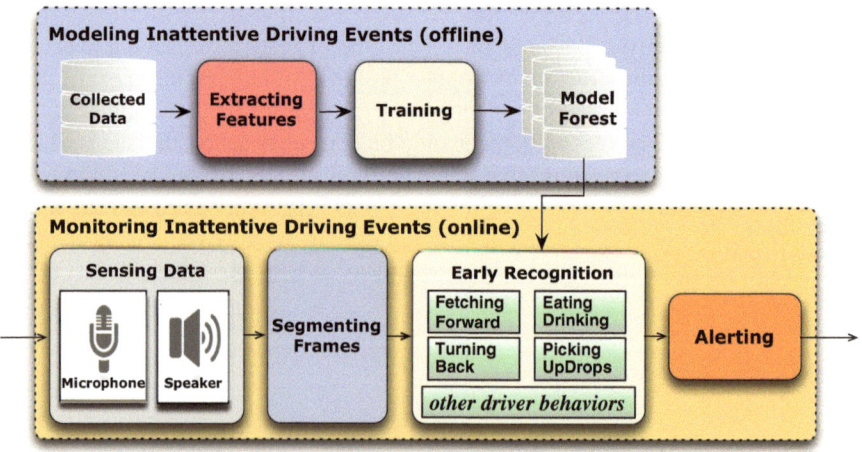

Fig. 4.3 System architecture and work flows

segments the corresponding frequency-time Doppler profile from the beginning to current time and sends to *Early Recognition* until ER outputs a recognition result or detects the end of the event. Then in *Early Recognition*, ER extracts features from segments and identifies whether the events are inattentive driving events or other driving behaviors at some early stages based on the trained model forest. Finally, if any of the four typical inattentive driving events are recognized through the above procedure, ER sends a warning message to alert driver, which is in the form of several short tones. Specially, the tones are set at about 1 kHz to make it clear enough to be heard, but not annoying to drivers.

4.3.2   Model Training at Offline Stage

4.3.2.1   Establishing Training Dataset

To collect data in real environments, we develop an Android-based program to generate and collect audio signals, and then transform the raw sampled signals to the frequency-time Doppler profiles.

We collect these transformed data from eight drivers with distinct vehicles. Eight smartphones of four different types are used, which are HTC Desire G7, ZTE U809, HTC EVO 3D and SAMSUNG Nexus5, two for each type. Meanwhile, all vehicles are equipped with cameras so that drivers' events can be recorded as the ground truth. Our data collection spans from October 23, 2015 to January 27, 2016, during which all the daily driving including commuting to work, shopping, touring, etc. is recorded. Drivers are not told about our purpose so that they can drive in a natural way. And each of our volunteer has their own driving routes differs from each other. After that, we ask five experienced drivers to watch the videos recorded by the cameras and recognize all types of inattentive driving events from the 3-month traces. In total, we obtain 3532 samples of inattentive driving events from the collected traces, which are severed as the ground truth. Afterwards, we combined the collected Doppler profiles of audio signals and their labels into a training dataset X.

4.3.2.2   Extracting Effective Features

Traditional feature extracting methods extract features by observing the unique patterns manually. Features extracted by these methods usually have redundant information and are poor in robustness. To achieve better features, ER leverages *Principal Components Analysis* (*PCA*) algorithm to the raw data.

In PCA algorithm, to extract features from training dataset X, a projection matrix W that contains feature vectors ranked by variance, is calculated using *Singular Value Decomposition* (*SVD*), which is given by $X = U \Sigma W^T$. For $m \times n$ matrix X, U is a $m \times m$ unitary matrix, Σ is a $m \times n$ matrix with non-negative singular values on the diagonal. W is a $n \times n$ unitary matrix, which has n orthogonal features ranked by importance. Since too many features may bring in the danger of over-fitting, we should select the minimum number of features, d, which contains enough information of the raw data. Considering the reconstruction property of PCA, the object function is

$$\min_{d} \ (\sum_{i=1}^{d} \sigma_i)/(\sum_{i=1}^{n} \sigma_i) \geq t \quad t \in [0, 1], \tag{4.2}$$

where σ_i is the ith largest singular value of matrix X, which denotes the importance of the ith features in W, and t is the threshold of reconstruction, denoting the remaining information of the raw data. In ER, t is set to be 0.95 to guarantee the features' validity. For all four inattentive driving events, we have $d = 17$ from Eq. (4.2).

However, considering the limited computational abilities of smartphones, $d = 17$ is slightly large for effective calculation. To further reduce d, we analyze inattentive driving events pairwise. According to Eq. (4.2), for any pair of the four inattentive driving events, if t is set to be 0.9, then we obtain $d = 2$. This result shows that if we compare the inattentive driving events pairwise, then $d = 2$ is good enough to represent most information of the raw data.

Figure 4.4 shows the distributions of Fetching Forward events versus other three inattentive driving events and normal driving in two-dimensional feature spaces. It is can be seen from Fig. 4.4 that Fetching Forward events can be discriminated from other inattentive driving events along with normal driving using two features extracted by PCA. Similarly, for all pairs for inattentive driving events along with normal driving, this conclusion remains. Therefore, in order to reduce the amount of features and improve recognition accuracy, we extract features for inattentive driving events pairwise.

4.3.2.3   Training a Multi-Classifier

After features extracting through PCA, we first use *Support Vector Machine (SVM)* with RBF kernel to train binary classifiers for every pair of inattentive driving events, as SVM with the RBF kernel can achieve relatively high performance across most training data size for binary classification task. In practice, we apply LibSVM [26] to implement our SVM algorithm. Based on the binary classifiers, a voting mechanism is proposed to form a multi-classifier to differentiate all four inattentive driving events.

Given that each binary classifier has one vote $v \in \{0, 1\}$ for building the multi-classifier. Considering a binary classifier for separating event a from event b, then for a specific event e, if the binary classifier identifies e as event a, then event a get a vote $v_a = 1$, event b get a vote $v_b = 0$. Assuming an event set E containing k types

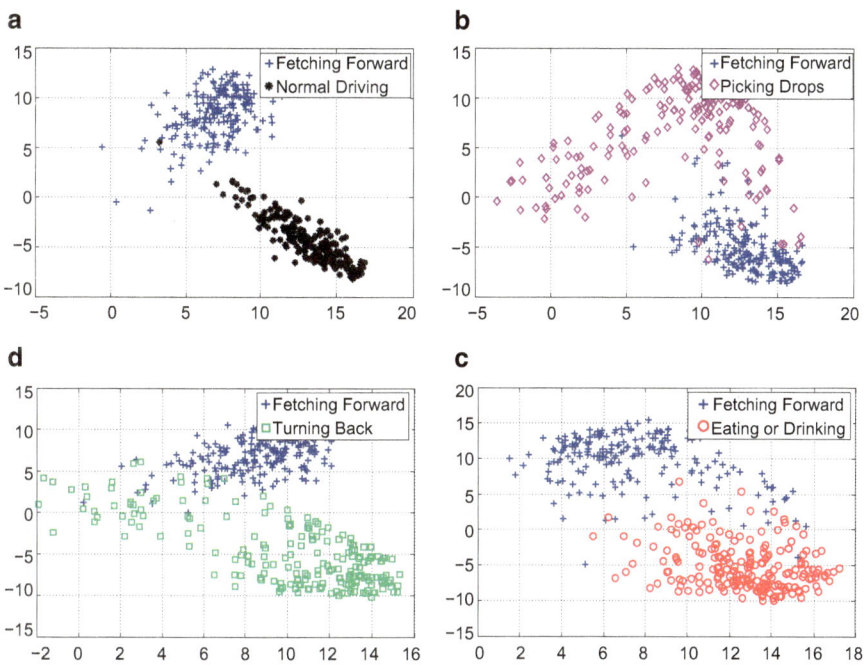

Fig. 4.4 Distributions of Fetching Forward events with other inattentive driving events and normal driving in two-dimensional feature space. (**a**) Fetching Forward vs. Normal Driving. (**b**) Fetching Forward vs. Picking up Drops. (**c**) Fetching Forward vs. Turning Back. (**d**) Fetching Forward vs. Eating & Drinking

of events, a classifier group has C_k^2 binary classifiers. For the event e, the votes of all C_k^2 binary classifiers can be denoted as

$$V(e) = \sum_{j \in [1, C_k^2]} v_j, \tag{4.3}$$

where v_j is a vote vector of k elements that denotes the vote of the jth binary classifier. The event class which get the most votes in $V(e)$, i.e.,

$$c = \max_j V_j(e) \quad j \in [1, k], \tag{4.4}$$

is supposed to the classified event of e.

Moreover, for a specific type of inattentive driving events, there are exactly $k - 1$ binary classifiers directly related to this type of event in all C_k^2 binary classifiers. So the votes of the winning event class should satisfy

$$V_c(e) = k - 1. \tag{4.5}$$

For an event which gets through the multi-classifier and gets a classification result c from Eq. (4.4), if it does not satisfy Eq. (4.5), ER considers the event as *other driving events*.

4.3.2.4   Setting Up Gradient Model Forest for Early Recognition

To approach the goal of recognizing inattentive driving events as early as possible, we propose an early recognition method.

Considering an inattentive driving events set E containing k types of events, $E = \{e_1, e_2, \cdots, e_k\}$. For a given event e started at t_0 and finished at t_1, the *completion degree* α of the event e at time t is computed as

$$\alpha_e = \frac{t - t_0}{t_1 - t_0} = \frac{\tau}{T} \quad t \in [t_0, t_1], \tag{4.6}$$

where τ denotes the *time duration* of e at time t and T denotes the *total length* of e. Obviously, $\alpha_e \in [0, 1]$. And when $\alpha_e = 1$, the event e finishes. Equation (4.6) shows that the goal to recognize an inattentive driving event e as early as possible is equivalent to finish recognition when α_e is as small as possible. As a result, based on different α of inattentive driving events at different τ, we set up a group of classifiers for early recognition, i.e. the gradient model forest.

For modeling the complete degree α at different time duration τ of inattentive driving events, the variation of the total length T among all events should first be considered. For different types of inattentive driving events, T varies because of the nature differences of the events, thus we set different models for different types of events. Moreover, for a specific type of inattentive driving events, T also varies depending on different drivers and driving situations. We also need to take this variation into consideration when setting models.

According to statistics of the dataset established in Sect. 4.3.2.1, the total length T for each type of inattentive driving events approximately satisfies a Gaussian distribution. For example, T of Fetching Forward events approximately satisfies a Gaussian distribution of mean value $\mu = 4.38$ s and standard deviation $\sigma = 0.32$ s. Since two standard deviations from the mean account for 95.45% data in Gaussian distribution, we can think that about 95% of Fetching Forward events have T from 3.74 s $(\mu - 2\sigma)$ to 5.02 s $(\mu + 2\sigma)$. As shown in Fig. 4.5, based on the completion degree-time duration relations of Fetching Forward events having T equals to 3.74, 4.38 and 5.02 s, a quadratic curve is fit to model the relationship between α and τ for Fetching Forward event, which starts at the origin, goes through the mid-point of the line $T = 4.38$ s and ends at the end of the line $T = 5.02$ s. For any $\tau > 5.02$ s, $\alpha = 1$. The fitting curve can thus represent most Fetching Forward events because it closes to most Fetching Forward events at some time period. With the similar analysis, ER models a relationship between α and τ for each type of inattentive driving events. Figure 4.6 shows relationships between α and τ for all four types of inattentive driving events.

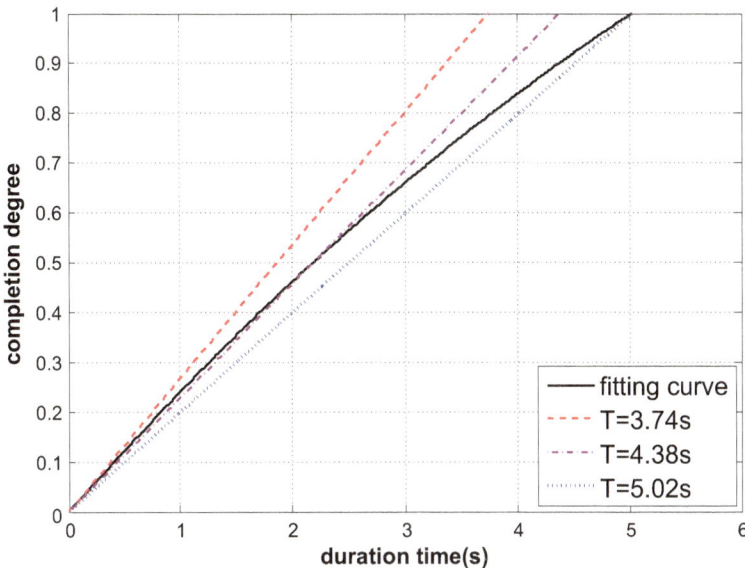

Fig. 4.5 The relationships between the completion degree α and time duration τ of fetching forward events under different the complete time T

With the relationships between α and τ, for any given time duration τ, ER can get a completion degree set $A^\tau = \{\alpha_1^\tau, \alpha_2^\tau, \cdots, \alpha_k^\tau\}$, which contains the completion degree for each type of inattentive driving events at time duration τ, as shown in Fig. 4.6. According to A^τ, ER segments the Doppler profiles of all types of inattentive driving events $X = \{X_1, X_2, \cdots, X_k\}$ and then gets the new input dataset $X^\tau = \{X_1^\tau, X_2^\tau, \cdots, X_k^\tau\}$. Selecting n different τ by gradient, we form a n-element time duration set $T = \{\tau_1, \tau_2, \cdots, \tau_n\}$. ER then segments the Doppler profiles based on T and ends up with a gradient dataset forest $X = \{X^{\tau_1}, X^{\tau_2}, \cdots, X^{\tau_n}\}$. Afterwards, X is trained through the methods in Sects. 4.3.2.2 and 4.3.2.3. Although for a specific dataset X^τ, patterns for parts of inattentive driving events are not guaranteed to be unique, ER can always get a multi-classifier θ^τ. Based on the new input dataset X, a gradient model forest $\Theta = \{\theta^{\tau_1}, \theta^{\tau_2}, \cdots, \theta^{\tau_n}\}$ is set up, and each of θ^τ is a matrix containing all binary classifiers as $\theta^\tau = ((\theta_1^\tau)^T; (\theta_2^\tau)^T; \ldots; (\theta_m^\tau)^T)$, where $m = C_k^2$ for all different pairwise inattentive driving events. Specially, the last multi-classifier of the model forest, i.e., θ^{τ_n}, is a multi-classifier for recognizing inattentive driving events after they finished. Finally, we obtain a gradient model forest Θ, which could be used to realize early recognition of inattentive driving events.

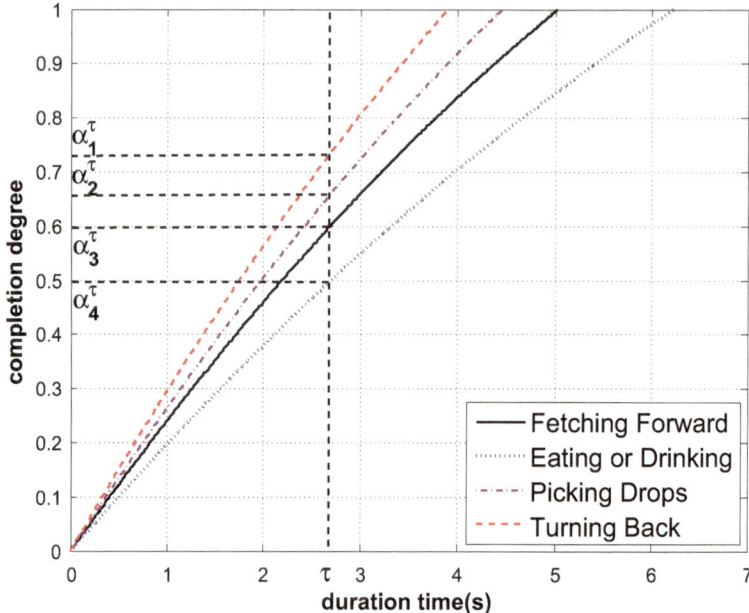

Fig. 4.6 The relationships between the completion degree α and time duration τ for four kinds of inattentive driving events

4.3.3   Recognizing Inattentive Driving Events at Online Stage

4.3.3.1   Segmenting Frames Through Sliding Window

In order to recognize current driving events, ER first needs to determine the time duration by recognizing the beginning and the end of the driving events.

As mentioned in Sect. 4.2.2, all driving events occur with positive and negative Doppler shifts in the frequency-time Doppler profiles, i.e., energy fluctuation near the pilot frequency (20 kHz) as shown in Fig. 4.2. From analyzing the traces collected in real driving environments, we find that when events occur, the average amplitude of frequency bands beside the pilot frequency keeps a relatively high value. Figure 4.7 shows the average amplitude of frequency bands beside the pilot frequency during 20 s driving containing a fetching forward event. From Fig. 4.7, it can be seen that the average amplitude for events is much greater than that without events.

Based on patterns of the average amplitude, ER employs the sliding window method to capture Doppler shifts caused by driving events. ER keeps computing the average amplitude within a window and compares with thresholds to determine the start point and end point of an event. The window size and thresholds can be learned from the collected data. When the start point of an event is detected, ER

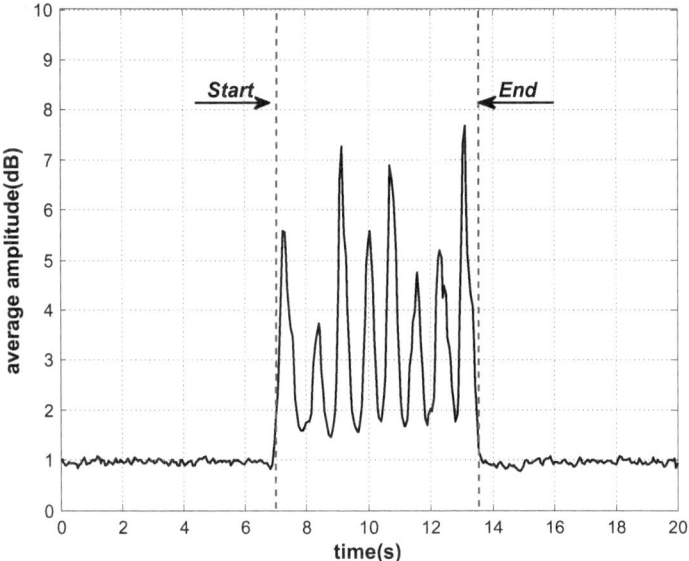

Fig. 4.7 The average amplitude of frequency bands except the pilot frequency during a 20 s driving containing a fetching forward event

segments a frame from the start point to current point and sends the frame to go through early recognition. ER keeps segmenting and sending at short time intervals until a recognition result is output or the end of the event is detected.

4.3.3.2   Detecting Inattentive Driving Events at Early Stage

After getting a frame of a driving event, according to the time duration of the frame, ER inputs the frame into the corresponding classifier in the model forest Θ to recognize the specific type of driving event. However, for frames with small time durations, the recognition results may not be accurate because these frames contains few information of the events. Thus, ER proposes a mechanism to guarantee the validity of early recognition.

Specifically, for a Doppler profile frame e of time duration $\tau \in [\tau_i, \tau_{i+1}]$, ER calls the classifier $\theta^{(\tau_i)}$ and $\theta^{(\tau_{i+1})}$ to recognize the driving event. From the two classifiers, ER gets the classification results c_1 and c_2. Then ER checks the value of c_1 and c_2 an d only when $c_1 = c_2$, ER admits the validity of the result and temporarily stores it. After ER detects several continuous valid results that denote the same inattentive driving event, the frame is confirmed as the specific type of inattentive driving event and an alert is sent to the driver. During the procedure, the number of continuous valid results which output the same is defined as *Convince Length*. The procedure continues until ER detects the end of the event, and if ER does not output a result before recognizing the end of an event, it uses the last classifier of the model forest, θ^{τ_n}, to finish the recognizing and get the corresponding output. Note that θ^{τ_n} leverages the full information of the event in Doppler profile, so it is equivalent to a classifier to detect inattentive driving event after the event is finished. For each event, ER records the output recognition result and the result from the classifier θ^{τ_n}. The proposed mechanism of ER can reduce mistaken recognition and avoid disturbing drivers from false warning effectively.

4.4 Evaluation

In this section, we first present the prototype of ER, then evaluate the performance of ER in real driving environments.

4.4.1 Setup

We implement ER as an Android App and install it on eight smartphones of four different types, which are HTC Desire G7, ZTE U809, HTC EVO 3D and SAMSUNG Nexus5, each type has two smartphones. Then ER is running by eight drivers with distinct vehicles in real driving environments to collect traces for evaluation. Drivers are not told about our purpose so that they can drive in a natural way. Meanwhile, each car is implemented with a camera for recording driver's driving behaviors and five experienced drivers are asked to recognize inattentive driving events as the ground truth. After data collection from March 11 to May 6, 2016 using method described in Sect. 4.3.2.1, during which all the daily driving including commuting to work, shopping, touring, etc., we obtain a test set with 1473 inattentive driving events to evaluate the performance of ER.

4.4.2 Metrics

To evaluate the performance of ER, we define metrics as follows, note that ρ_{ij} denotes the situation that the event of ground truth j is recognized by ER system as event i.

- *Accuracy*: The probability that an event is correctly identified for all K types of events, i.e., $Accuracy = \frac{\sum_{i=1}^{K} \rho_{ii}}{\sum_{j=1}^{K} \sum_{i=1}^{K} \rho_{ij}}$.

- *Precision*: The probability that the identification for an event A is exactly A in ground truth, i.e., $Precision_k = \frac{\rho_{kk}}{\sum_{i=1}^{K} \rho_{ik}}$.
- *Recall*: The probability that an event A in ground truth is identified as A, i.e., $Recall_k = \frac{\rho_{kk}}{\sum_{j=1}^{K} \rho_{kj}}$.
- *False Positive Rate* (*FPR*): The probability that an event not of type A is identified as A, i.e., $FPR_k = \frac{\sum_{j=1}^{K} \rho_{kj} - \rho_{kk}}{\sum_{j=1}^{K} \sum_{i=1}^{K} \rho_{ij} - \sum_{i=1}^{K} \rho_{ik}}$.
- *F-Score*: A metric that combines precision and recall, i.e., $F\text{-}score_k = 2 \times \frac{Precision \times Recall}{Precision + Recall}$.

We use F-score as our major metric to evaluate the recognition accuracy in the following evaluation, because F-score is a metric that combines precision and recall, which can provide a more convincing result than precision or recall alone. Also, it is a more fine-grained result than accuracy.

4.4.3 Overall Performance

4.4.3.1 Total Accuracy

Figure 4.8 plots the recognition accuracy of ER system and the original SVM classifier θ^{τ_n} for eight drivers, it can be seen that ER achieves a total accuracy of 94.80% for recognizing all types of inattentive driving events, while the total accuracy for θ^{τ_n} is 84.78%. Further, ER performs far better than the original SVM classifier θ^{τ_n} for any of the eight drivers. The lowest accuracy for ER of the eight drivers is 91.73%, which validate the effectiveness and stability of ER in real driving environments.

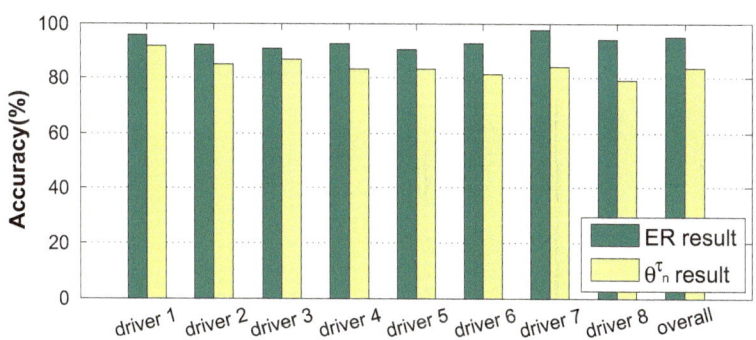

Fig. 4.8 The total accuracy of ER and classifier model θ^{τ_n} over eight drivers

4.4.3.2   Recognizing Inattentive Driving Events

For different types of inattentive driving events, the precision, recall and F-score for recognition is showed in Fig. 4.9a. It can be seen that all these three metrics are high for every type of inattentive driving events. Specifically, the precision is no less than 89%, while the recall is above 91%, and the F-score is more than 92%. Further, ER has a better performance with all three metrics above 95% when recognizing *Fetching Forward* and *Eating or drinking* events since their patterns are more distinctive as shown in Fig. 4.2. For sensing *Turning Back* and *Picking Drops*, the performance drops a little due to the characters of these events, but still achieves more than 92% in F-score. As a result, ER could recognize all four different types of inattentive driving events with high accuracy.

Moreover, for each of the eight drivers, we evaluate the FPRs of recognizing specific type of inattentive driving events. Figure 4.9b shows the box-plot of the FPRs for each type of inattentive driving events. We can observe from Fig. 4.9b that the highest FPR is no more than 2.5% and the average FPR is as low as 1.4% over the four events and eight drivers. Moreover, the FPR for sensing *Picking Drops* is lower than 1%, which is the best among four types of inattentive driving events. This is because the motion of *Picking Drops* is more robust than other event. It shows that ER could realize inattentive driving events recognition with few false alarms, which is user-friendly for drivers.

4.4.3.3   Realizing Early Recognition

We plot the Cumulative Density Function (CDF) of recognition time for each type of inattentive driving events and the CDF of all types of events in Fig. 4.10. It can be

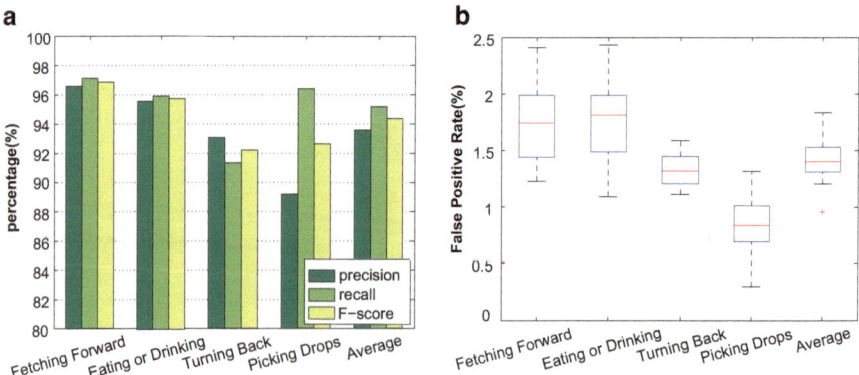

Fig. 4.9 The recognition results on detecting inattentive events. (**a**) The precision, recall and F-Score for all types of inattentive events. (**b**) Box plot of False Positive Rate of all types of inattentive events over eight drivers

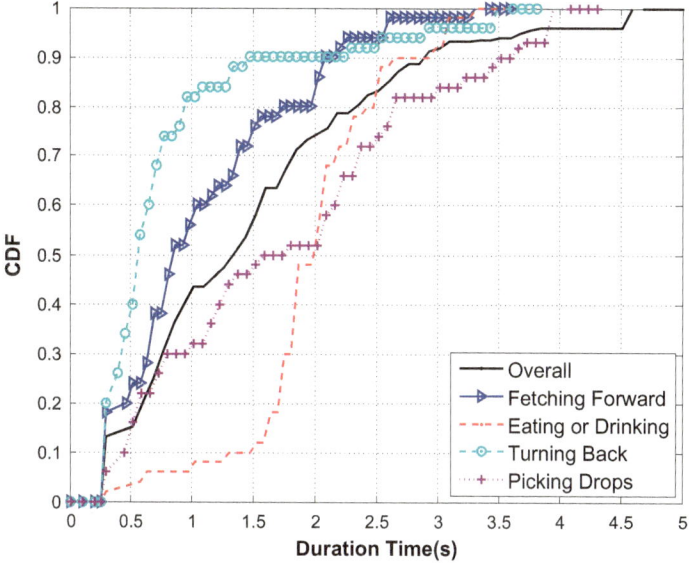

Fig. 4.10 The CDF of recognition time for all types of inattentive events

seen from Fig. 4.10 that 50% of all inattentive driving events are recognized by ER before 1.4 s and 80% can be recognized before 2.3 s, while the average total length of all events is around 4.6 s. In another word, more than 80% inattentive driving events can be recognized at the time less than 50% of the average total length of all events. For each specific type of events, the 80%-recognized time are around 2, 2.5, 1.0 and 2.6 s for Fetching Forward, Eating or drinking, Turning Back and Picking Drops respectively. And the corresponding average total length for the four events are 4.3, 5.4, 3.5 and 4.0 s, which validate the early recognition property of ER system.

From the overall performance of ER, we see ER not only realize effective early recognition, but further achieve much higher recognition accuracy than the original SVM multi-classifier.

4.4.4   Impact of Training Set Size

According to Sect. 4.3.2.1, we collect 3532 inattentive driving events in total for training. Figure 4.11 shows the impact of training set size. It can be seen that the F-score rises as training set size increases and goes stable after a certain size for each inattentive driving events. Specifically, to get a stable F-score, ER needs at least 220 training samples for training Fetching Forward events, 300 samples for Eating or

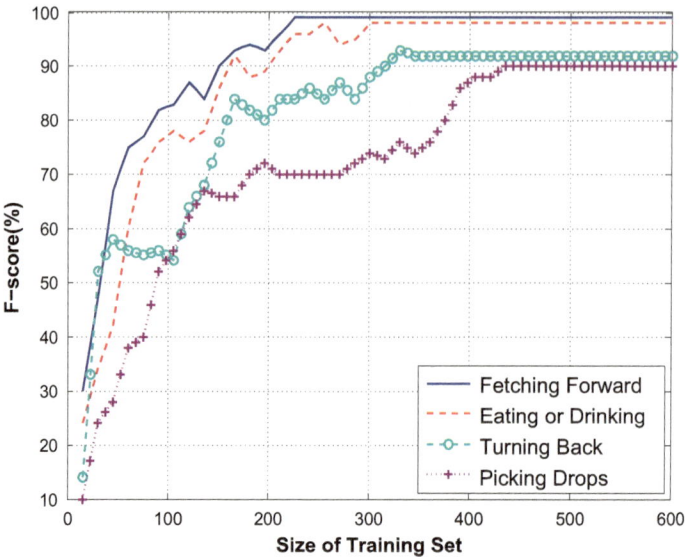

Fig. 4.11   F-score under different size of training set for all types of inattentive events

Drinking, 340 for Turning Back and 450 for Picking Drops. In practice, we use as much training examples as we can get to guarantee the performance of ER.

4.4.5   Impact of Road Types and Traffic Conditions

Different road types and traffic conditions may influent drivers' driving behaviors and vehicle conditions, thus may have impacts on the performance of ER. We analyze the collected traces of different road types (local road and highway) and different traffic conditions (during peak time and off-peak time), respectively. Figure 4.12 shows the result. It can be seen that ER achieves fairly good F-scores for recognition at any combination of road types and traffic conditions. In addition, during peak time, the F-scores of ER is slightly lower than the F-scores during off-peak time because heavy traffic condition may bring more stops for vehicle and more driving behaviors such as shifting gears, which may result in more mistaken recognitions. Further, the F-scores of ER when driving on highway is slightly higher than the F-scores on local road since drivers are more concentrate when driving at high speed and the road on highway is more smooth than local road, which brings less influence to ER.

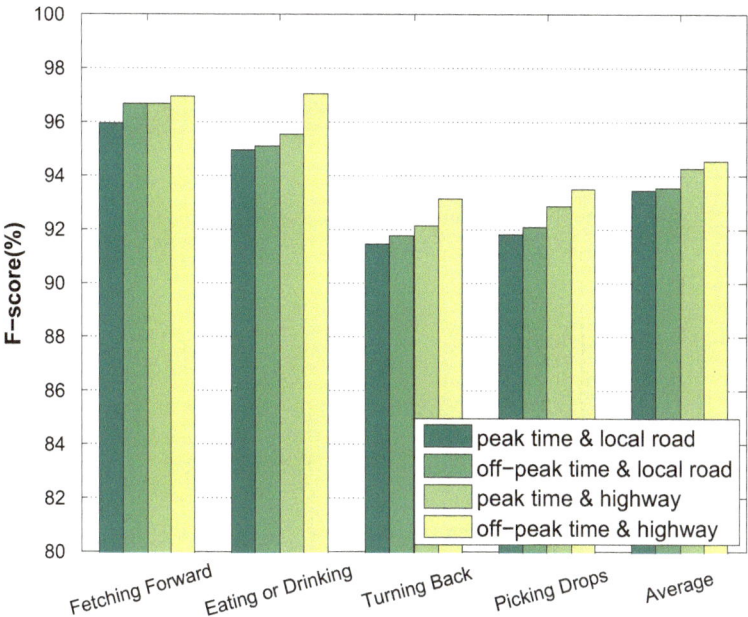

Fig. 4.12   F-score under different traffic conditions and road types for all inattentive events

4.4.6   *Impact of Smartphone Placement*

In our experiments, each driver place the smartphone randomly on instrument panel (left side), instrument panel (middle part), panel near cab door and cup-holder, or in driver's pocket for daily driving. Figure 4.13 shows that ER can achieve fairly good F-scores for recognitions under different smartphone placements. Specifically, smartphones placed on instrument panel achieve best recognition results as the audio devices of smartphones is directly face to drivers. And smartphones placed in drivers' pockets achieves lower F-score than others because the movement of drivers may bring influence to ER. But the F-score for any smartphone placement and any inattentive driving events is above 88%, which is acceptable for using ER in real driving environments.

4.5   Conclusion

In this chapter, we address the problem of recognizing inattentive driving as early as possible to improve driving safety. In particular, we propose an early recognition system, ER, to recognize different inattentive driving events at the early stage leveraging build-in audio devices on smartphones. To achieve the recognition,

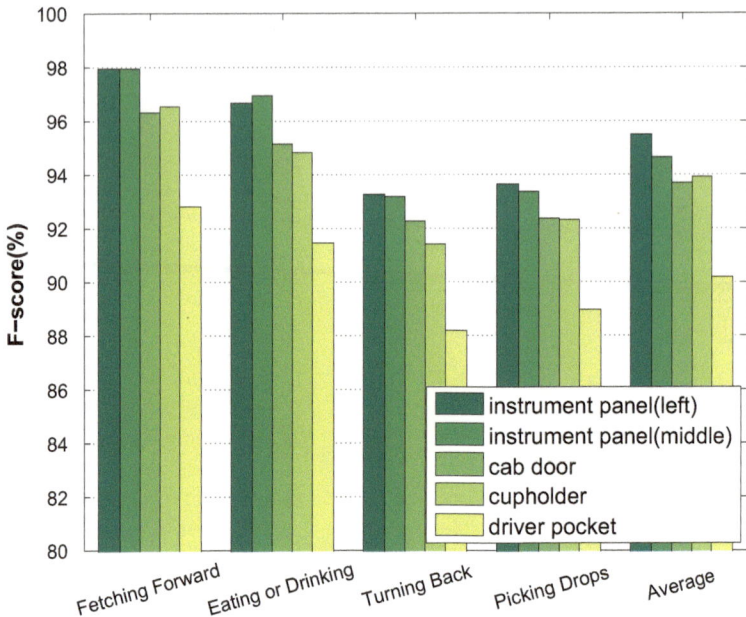

Fig. 4.13 F-score under different smartphone placements for all types of inattentive events

ER first extracts effective features using PCA based on the patterns of different inattentive driving events on Doppler profiles of audio signals. Then ER leverages SVM to form a multi-classifier and further set up a gradient model forest for early recognition. The extensive experiments in real driving environments show that ER achieves high accuracy for inattentive driving recognition and realizes recognizing events at early stage.

Chapter 5
State-of-Art Researches

5.1 Smartphone Sensing Researches

Smartphone sensing has continuous been a hot issue for researchers in recent decades. Works involving smartphone sensing nowadays covers human activity recognition [36, 37], context recognition [38, 39], social network analyzing [40], environmental monitoring [41, 42], health-care [43], smart transportation systems [7], etc. As smartphone sensing been so large an research area, we further divide smartphone sensing researchers into several categories based on the main sensors embedded in smartphones.

Cameras in smartphones are most widely exploited for smartphone sensing systems and applications. These camera-based researches includes traditional tasks such as photo or video analysis [44, 45] to more specialized sensing activities such as tracking the user's eye movement to activate applications in the phone [46, 47]. Moreover, cameras in smartphones are utilized to more areas nowadays. Werner et al. [48] propose a distance estimation algorithm to achieve indoor localization using the camera of a mobile device. Kwon et al. [49] show that video data captured by smartphone camera can be exploited to extract the heartbeat of people. More recently, Boubezari et al. [50] propose a method to leverage smartphone cameras for communication through visible light.

Besides camera-based smartphone sensing researches, there have been active research works in exploiting motion sensors in smartphones for sensing tasks. Harrison et al. [36] and Bao and Intille [37] leverage accelerometer as well as gyroscope on a smartphone to sense human activity in daily lives. Miluzzo et al. [51] show that accelerometer data is capable of characterizing the physical movements of the user carrying the phone. Liang and Wang [52] leverage accelerometer data as well as a stream of location estimates from GPS to recognize the mode of transportation of a user, such as using a bike, car, taking a bus, or the subway.

© The Author(s), under exclusive licence to Springer International Publishing AG, part of Springer Nature 2018
J. Yu et al., *Sensing Vehicle Conditions for Detecting Driving Behaviors*,
SpringerBriefs in Electrical and Computer Engineering,
https://doi.org/10.1007/978-3-319-89770-7_5

Environment sensors, such as barometers, hygrometers and thermometers, also play an important role in developing smartphone sensing applications and systems, especially in environment monitoring [41, 42]. Honicky et al. [41] study air quality and pollution using specialized sensors embedded in prototype mobile phones. Min et al. [42] uses sensors in phones to build a system that enables personalized environmental impact reports, which track how the actions of individuals contribution to the changes of environment such as carbon emissions. These tasks often require mobile sensing methods as well as crowd sensing technologies, for instance, building a noise map of a city [53].

More recently, technologies are developed rapidly of smartphone sensing leveraging acoustic devices on smartphones. ApneaApp [54] uses FMCW to track heartbeats by analyzing periodic patterns of acoustic signals. AAmouse [55] realizes a virtual mouse leveraging audio devices of smartphones. CAT [56], LLAP [57], FingerIO [58] and Strata [59] track movements of a finger or a hand using a smartphone to transmitting and receiving acoustic signals, all these methods can achieve sub-centimeter tracking error.

5.2 Vehicle Dynamics Sensing Researches

For sensing basic vehicle dynamics, GPS embedded in vehicles is most commonly used. Since GPS may suffer from low accuracy in urban environments, there has been many progresses on enhancing the accuracy of GPS, such as DGPS, WAAS, and European Geostationary Navigation Overlay Service (EGNOS). Differential GPS (DGPS) uses a network of fixed reference stations to broadcast the differences between positions from GPS and known ones [60]. Wide Area Augmentation System (WAAS) uses reference stations to measure the correction and broadcast the messages [61]. These augmentations can improve the accuracy of GPS from about 15 m to less than 5 m. More recently, several high-precision GPS receivers [62, 63] are designed to give fine-grained localization solutions. However, a national construction of such reference stations is still costly. With the development of Internet of Things (IoT) in the last decade, Vehicle-to-Vehicle (V2V) communication, as a method for sensing basic vehicle dynamics, has drawn the attention of researchers. Based on Bayesian model, Rohani et al. [64] sense vehicle dynamics by sharing GPS data and inter-vehicular distance measurements. Bento et al. [65] propose an inter-vehicle multi-sensor fusion supported by both V2V and V2I (vehicle-to-infrastructure) for sensing vehicle dynamics. Li et al. [66] merge the local maps of several vehicles to indirectly estimate the positions. Hedgecock et al. [67] use GPS from pairs of objects to directly calculate their relative coordinate and test it in a lane-change situation. Golestan et al. [68] measure the belief of each vehicle about its dynamics, then use V2V and Particle Filter to incorporate those beliefs. All these methods show satisfactory results, but they are hard to implement because they need to establish a robust communication network environment among the vehicles, which is still not widely achieved among vehicles today.

For sensing vehicle speed, OBD-II adapter [3] is a popular interface to provide the vehicle speed in real-time. Acoustic wave sensors [69, 70] are utilized to estimate the vehicle speed in open environments. Furthermore, traffic magnetic sensors are also employed to capture the vehicle speed [71]. These approaches need to install additional hardware to perform speed estimation. To eliminate the need of pre-deployed infrastructures and additional hardware, recent studies concentrate on using cell phones to measure the vehicle speed. In particular, Hoh et al. [72] and Thiagarajan et al. [73] use GPS or sub-sampled GPS embedded in smartphones to drive the vehicle speed. Although GPS is a simple way to obtain vehicle speed, the urban canyon environment and the low update frequency of GPS make it difficult to accurately capture the frequent changing vehicle speed in urban environments. And continuously using GPS causes quicker battery drainage on smartphones. Knowing the drawbacks of using GPS, Chandrasekaran et al. [74, 75] estimate the vehicle speed by warping mobile phone signal strengths and Gundlegard and Karlsson [76] and Thajchayapong et al. [77] use the handovers between base stations to measure the vehicle speed. These solutions need to build a signal database which may incur high cost and cannot achieve high estimation accuracy.

5.3 Driver Behaviors Detection Researches

For researches on driver behaviors detection, some existing works realize driver behaviors detection by using professional infrastructure including EEG [16] and water cluster detectors [78], or common sensors such as infrared sensors [79] and cameras [17]. Specifically, Yeo et al. [16] use an EGG equipment which samples the driver's EGG signals to detect drowsiness during car driving. Sakairi and Togami [78] couple a breath sensor with an alcohol sensor, and simultaneously detects the electrical signals of both breath and alcohol in the breath for detecting driving fatigues as well as drunk driving. Lee et al. [79] use infrared sensors monitoring the driver's head movement to detect drowsy driving. In [17], GPS, cameras, alcohol sensor and accelerometer sensor are used to detect driver's status of drunk, fatigued, or reckless. However, the solutions all rely on pre-deployed infrastructure and additional hardware that incur installation cost. Moreover, those additional hardware could suffer the difference of day and night, bad weather condition and high maintenance cost.

To overcome the limitations of pre-deployed infrastructure, recent studies put their efforts to exploit mobile devices on driver behaviors detection, which can be categorized as vision-based solutions [31, 80] and sensor-based solutions [3, 22, 81]. In vision-based solutions, the build-in cameras are used to capture the graphic information for processing. You et al. [80] leverage dual cameras of smartphones to tract road conditions and detect drivers' status at the same time. However, the accuracy of vision-based approaches is unstable depends on weather, lighting and smartphones placement. In sensor-based solutions, Chen et al. [81] combine sensors by using Inertial Measurement Units on smartphones to detect various steering maneuvers of a driver. This approach can only provide detection results after driving behaviors finished. Besides, Wang et al. [3] use accelerators of smartphones to determine usage of phones while driving, but this work cannot recognize other driving behaviors but usage of phones. Further, there are several work focusing on sensing steer wheel motions for detecting drivers' operations with the vehicle [82, 83], Lawoyin et al. [82] estimate the rotation angles of steering wheel utilizing steering-wheel-mounted sensors. Karatas et al. [83] present a method to track the rotation angle of steering wheel based on smart-watches.

5.4   Common Issues

For the above state-of-art researches, there are still several common issues, which are listed as following:

- For researches of sensing vehicle dynamics, the existing works either have poor accuracy (especially GPS-based approaches), or need extra pre-implemented devices (e.g., OBD adapter) and infrastructures (for methods related to V2V and V2I). However, there are few approaches providing light-weight solutions that can achieve high accuracy for sensing vehicle dynamics. Unlike these works, we propose smartphone-based approaches for accurately sensing varies vehicle dynamics as described in Chap. 2.
- For researches of detecting vehicle behaviors, besides pre-implemented devices and infrastructures (mostly special designed sensors), there have been many existing researches utilizing mobile devices to perform the detection. However, most works with mobile devices can only provide coarse-grained results (i.e., abnormal vs normal), and fail to provide fine-grained results (specific types of vehicle behaviors). Unlike these works, we propose a fine-grained vehicle behaviors detection system based on smartphones as described in Chap. 3.
- For researches of detecting driver behaviors, existing works focus on detecting different types of driver behavior, i.e., drunk, fatigue, drowsy, etc. Many approaches achieve high performance for detecting specific types of driver behaviors. However, none of these works can provide detections results at the early stage when drivers are performing certain behaviors. Unlike these works, we propose a early recognition system for detecting inattentive driving behaviors as described in Chap. 4.

Chapter 6
Summary

6.1 Conclusion of the Book

To achieve the goal of safe and convenient driving, in this book, we dig into the problem of explore smartphone-based approaches and systems for sensing vehicle conditions and further detect driving behaviors. Specifically, in Chap. 2, we address the problem of enhancing off-the-shelf smartphone for sensing vehicle dynamics to support vehicular applications. Two smartphone sensors, i.e., accelerometer and gyroscope, are exploited to sensing five different types of vehicle dynamics, i.e., the moving and stopping, driving on uneven road, turning, changing lanes and the instance speed of vehicles when driving.

In Chap. 3, we focus on the problem of detecting and identifying abnormal driving behaviors of vehicles to improve driving safety. A smartphone-based system, D^3, is proposed to detect and identify specific types of abnormal driving behaviors by sensing the vehicle's acceleration and orientation using smartphone sensors. By training a multi-class classifier model through SVM based on the acceleration and orientation patterns of specific types of driving behaviors, D^3 not only implements coarse-grained detections but also conducts fine-grained identifications.

In Chap. 4, we deal with the problem of recognizing inattentive driving of drivers as early as possible to improve driving safety. In particular, an early recognition system, ER, is proposed to recognize different inattentive driving events at the early stage leveraging build-in audio devices on smartphones. Based on the effective features extracted using PCA from the patterns of different inattentive driving events on Doppler profiles of audio signals, a gradient model forest is proposed for early recognition of specific types of inattentive driving events.

From all the proposed methods in this book (described in Chaps. 2–4), we conduct month-level data collection through multiple voluntary drivers in real driving environments to get the data for analysis. Further, the proposed approaches

J. Yu et al., *Sensing Vehicle Conditions for Detecting Driving Behaviors*, SpringerBriefs in Electrical and Computer Engineering, https://doi.org/10.1007/978-3-319-89770-7_6

are mostly implemented as Android Apps and go through another month-lever experiments in real driving environments for validate and further evaluation. Extensive experiments show that all of proposed approaches achieves high accuracy of sensing and robust enough in real driving environments.

6.2 Future Directions

Besides the proposed approaches, there are also challenges and issues for utilizing smartphones for sensing vehicle and driver behaviors, which need to be further studied in the future, including:

- **Sensing Data Privacy**: When analyzing sensor readings on smartphones, it is fundamental to protect the privacy of users, as the sensing data may contains private information about users and their vehicles. One way to meet this requirement is to operate as much as on the smartphone without exposing the raw data to the external world, and communicating only the processing result of the inference to external entities. However, since sometimes the computation procedure may offload to other devices through network, it is necessary to keep the sensor data not been stolen.
- **Sensing Data Communication**: In this book, we focus on developing smartphone sensing approaches and systems based on driver's smartphone in a single vehicle, which cannot providing information beyond the scope of the vehicle. In practice, it is always useful to achieve information related to other vehicles and the whole driving environment. One possible way is to use *Vehicular Networks*. However, currently, the structure of *Vehicular Networks* may not be suitable for smartphone sensing data communication, which need to be supply in the future.
- **Sensing Data in Automated Driving**: Most recently, automated driving has been a hot research issue and draw the attention of many researchers from different areas. Typically, automated driving requires the vehicle itself can achieve basic information for maintaining safe driving. However, this information may not be accessible or even understandable for people in the vehicle. Therefore, it is reasonable to consider the automated driving-related scenarios and study the availability for applying the smartphone sensing to make the information more clear to people in the automated vehicle.

References

1. J. Levinson and S. Thrun, "Robust vehicle localization in urban environments using probabilistic maps," in *Proceedings of IEEE International Conference on Robotics and Automation (IEEE ICRA 2010)*, pp. 4372–4378, 2010.
2. F. Chausse, J. Laneurit, and R. Chapuis, "Vehicle localization on a digital map using particles filtering," in *Proceedings of IEEE Intelligent Vehicles Symposium (IEEE IV 2005)*, pp. 243–248, 2005.
3. Y. Wang, J. Yang, H. Liu, Y. Chen, M. Gruteser, and R. P. Martin, "Sensing vehicle dynamics for determining driver phone use," in *Proceeding of the 11th annual international conference on Mobile systems, applications, and services (ACM Mobisys 2013)*, pp. 41–54, 2013.
4. J. White, C. Thompson, H. Turner, B. Dougherty, and D. C. Schmidt, "Wreckwatch: Automatic traffic accident detection and notification with smartphones," *Mobile Networks & Applications*, vol. 16, no. 3, pp. 285–303, 2011.
5. J. Paefgen, F. Kehr, Y. Zhai, and F. Michahelles, "Driving behavior analysis with smartphones: insights from a controlled field study," in *Proceedings of the 11th International Conference on mobile and ubiquitous multimedia (ACM MUM 2012)*, pp. 36–43, 2012.
6. D. A. Johnson and M. M. Trivedi, "Driving style recognition using a smartphone as a sensor platform," in *Proceedings of IEEE International Conference on Intelligent Transportation Systems (IEEE ITSC 2011)*, pp. 1609–1615, 2011.
7. P. Mohan, V. N. Padmanabhan, and R. Ramjee, "Nericell: rich monitoring of road and traffic conditions using mobile smartphones," in *Proceedings of the 6th ACM conference on Embedded network sensor systems (ACM SenSys 2008)*, pp. 323–336, 2008.
8. T. N. Schoepflin and D. J. Dailey, "Dynamic camera calibration of roadside traffic management cameras for vehicle speed estimation," *IEEE Transactions on Intelligent Transportation Systems (IEEE TITS)*, vol. 4, no. 2, pp. 90–98, 2003.
9. Y. Wang, J. Yang, Y. Chen, H. Liu, M. Gruteser, and R. P. Martin, "Tracking human queues using single-point signal monitoring," in *Proceedings of the 12th annual international conference on Mobile systems, applications, and services (ACM MobiSys 2014)*, pp. 42–54, 2014.
10. Z. Wu, J. Li, J. Yu, Y. Zhu, G. Xue, and M. Li, "L3: Sensing driving conditions for vehicle lane-level localization on highways," in *Proceedings of IEEE Conference on Computer Communication (IEEE INFOCOM 2016)*, pp. 1–9, 2016.

J. Yu et al., *Sensing Vehicle Conditions for Detecting Driving Behaviors*, SpringerBriefs in Electrical and Computer Engineering, https://doi.org/10.1007/978-3-319-89770-7

11. X. Xu, J. Yu, Z. Yanmin, Z. Wu, J. Li, and M. Li, "Leveraging smartphones for vehicle lane-level localization on highways," *IEEE Transactions on Mobile Computing (IEEE TMC)*, 2017, doi:10.1109/TMC.2017.2776286.

12. H. Han, J. Yu, H. Zhu, Y. Chen, J. Yang, Y. Zhu, G. Xue, and M. Li, "Senspeed: Sensing driving conditions to estimate vehicle speed in urban environments," in *Proceedings of IEEE Conference on Computer Communications (IEEE INFOCOM 2014)*, pp. 727–735, 2014.

13. J. Yu, H. Zhu, H. Han, Y. Chen, J. Yang, Y. Zhu, Z. Chen, G. Xue, and M. Li, "Senspeed: Sensing driving conditions to estimate vehicle speed in urban environments," *IEEE Transactions on Mobile Computing (IEEE TMC)*, vol. 15, pp. 202–216, 2016.

14. World Health Organisation, "The top ten causes of death." [Online], Available: http://www.who.int/mediacentre/factsheets/fs310/en/, 2017.

15. C. Saiprasert and W. Pattara-Atikom, "Smartphone enabled dangerous driving report system," in *Proceedings of 46th Hawaii International Conference on System Sciences (IEEE HICSS 2013)*, pp. 1231–1237, 2013.

16. M. V. Yeo, X. Li, *et al.*, "Can SVM be used for automatic EEG detection of drowsiness during car driving?," *Safety Science*, vol. 47, no. 1, pp. 115–124, 2009.

17. S. Al-Sultan, A. H. Al-Bayatti, and H. Zedan, "Context-aware driver behavior detection system in intelligent transportation systems," *IEEE transactions on vehicular technology (IEEE TVT)*, vol. 62, no. 9, pp. 4264–4275, 2013.

18. S. Reddy, M. Mun, J. Burke, D. Estrin, M. Hansen, and M. Srivastava, "Using mobile phones to determine transportation modes," *ACM Transactions on Sensor Networks (ACM TOSN)*, vol. 6, no. 2, pp. 13–40, 2010.

19. J. Dai, J. Teng, *et al.*, "Mobile phone based drunk driving detection," in *Proceedings of EAI International Conference on Pervasive Computing Technologies for Healthcare (EAI PervasiveHealth 2010)*, pp. 1–8, 2010.

20. M. Fazeen, B. Gozick, R. Dantu, M. Bhukhiya, and M. C. González, "Safe driving using mobile phones," *IEEE Transactions on Intelligent Transportation Systems (IEEE TITS)*, vol. 13, no. 3, pp. 1462–1468, 2012.

21. U.S.NHTSA, "The visual detection of DWI motorists." [Online], Available: http://www.shippd.org/Alcohol/dwibooklet.pdf, 2015.

22. Z. Chen, J. Yu, Y. Zhu, Y. Chen, and M. Li, "D3: Abnormal driving behaviors detection and identification using smartphone sensors," in *Proceedings of 12th Annual IEEE International Conference on Sensing, Communication, and Networking (IEEE SECON 2015)*, pp. 524–532, 2015.

23. J. Yu, Z. Chen, Y. Zhu, Y. J. Chen, L. Kong, and M. Li, "Fine-grained abnormal driving behaviors detection and identification with smartphones," *IEEE Transactions on Mobile Computing (IEEE TMC)*, vol. 16, no. 8, pp. 2198–2212, 2017.

24. P. Harrington, *"Machine learning in action"*, vol. 5. Manning Greenwich, CT, 2012.

25. Y. Guo, L. Yang, X. Ding, J. Han, and Y. Liu, "Opensesame: Unlocking smart phone through handshaking biometrics," in *Proceedings of IEEE Conference on Computer Communication (IEEE INFOCOM 2013)*, pp. 365–369, 2013.

26. C. Chang and C. Lin, "Libsvm: A library for support vector machines," *ACM Transactions on Intelligent Systems and Technology (ACM TIST)*, vol. 2, no. 3, pp. 27–65, 2011.

27. USLegal, "Inattentive driving law and legal definition." [Online], Available: http://definitions.uslegal.com/i/inattentive-driving, 2016.

28. U. D. of Transportation, "Traffic safety facts research note, distracted driving 2014." [Online], Available: https://crashstats.nhtsa.dot.gov/Api/Public/ViewPublication/812260, 2016.

29. U. D. of Transportation, "Faces of distracted driving." [Online], Available: http://www.distraction.gov/faces/, 2016.

30. C. C. Liu, S. G. Hosking, and M. G. Lenné, "Predicting driver drowsiness using vehicle measures: Recent insights and future challenges," *Journal of safety research*, vol. 40, no. 4, pp. 239–245, 2009.

31. H. Dahlkamp, A. Kaehler, D. Stavens, S. Thrun, and G. R. Bradski, "Self-supervised monocular road detection in desert terrain," in *Proceedings of Robotics: science and systems (RSS 2006)*, pp. 1–7, 2006.

32. Q. Pu, S. Gupta, S. Gollakota, and S. Patel, "Whole-home gesture recognition using wireless signals," in *Proceedings of the 19th annual international conference on Mobile computing & networking (ACM MobiCom 2013)*, pp. 27–38, 2013.

33. X. Xu, H. Gao, J. Yu, Y. Chen, Y. Zhu, G. Xue, and M. Li, "Er: Early recognition of inattentive driving leveraging audio devices on smartphones," in *Proceedings of IEEE Conference on Computer Communications (IEEE INFOCOM 2017)*, pp. 1–9, 2017.

34. X. Xu, J. Yu, Y. Chen, Z. Yanmin, S. Qian, and M. Li, "Leveraging audio signals for early recognition of inattentive driving with smartphones," *IEEE Transactions on Mobile Computing (IEEE TMC)*, 2017, doi:10.1109/TMC.2017.2772253.

35. J. Yang, S. Sidhom, G. Chandrasekaran, T. Vu, H. Liu, N. Cecan, Y. Chen, M. Gruteser, and R. P. Martin, "Detecting driver phone use leveraging car speakers," in *Proceedings of the 17th annual international conference on Mobile computing and networking (ACM MobiCom 2011)*, pp. 97–108, 2011.

36. B. L. Harrison, S. Consolvo, and T. Choudhury, "Using multi-modal sensing for human activity modeling in the real world," in *Handbook of Ambient Intelligence and Smart Environments*, pp. 463–478, Springer, 2010.

37. L. Bao and S. Intille, "Activity recognition from user-annotated acceleration data," in *Proceedings of Second International Conference on Pervasive computing*, pp. 1–17, Springer, 2004.

38. B. Schilit, N. Adams, and R. Want, "Context-aware computing applications," in *Proceedings of First Workshop on Mobile Computing Systems and Applications (ACM HotMobile 1994)*, pp. 85–90, 1994.

39. H. W. Gellersen, A. Schmidt, and M. Beigl, "Multi-sensor context-awareness in mobile devices and smart artifacts," *Mobile Networks and Applications*, vol. 7, no. 5, pp. 341–351, 2002.

40. C. Intanagonwiwat, R. Govindan, and D. Estrin, "Directed diffusion: A scalable and robust communication paradigm for sensor networks," in *Proceedings of the 6th annual international conference on Mobile computing and networking (ACM MobiCom 2000)*, pp. 56–67, 2000.

41. R. Honicky, E. A. Brewer, E. Paulos, and R. White, "N-smarts: networked suite of mobile atmospheric real-time sensors," in *Proceedings of the second ACM SIGCOMM workshop on Networked systems for developing regions*, pp. 25–30, 2008.

42. M. Min, S. Reddy, K. Shilton, N. Yau, J. Burke, D. Estrin, M. Hansen, E. Howard, and R. West, "Peir, the personal environmental impact report, as a platform for participatory sensing systems research," in *Proceedings of the 7th International Conference on Mobile Systems, Applications, and Services (ACM MobiSys 2009)*, pp. 55–68, 2009.

43. M.-Z. Poh, K. Kim, A. D. Goessling, N. C. Swenson, and R. W. Picard, "Heartphones: Sensor earphones and mobile application for non-obtrusive health monitoring," in *Proceedings of International Symposium on Wearable Computers (IEEE ISWC 2009)*, pp. 153–154, 2009.

44. P. Peng, L. Shou, K. Chen, G. Chen, and S. Wu, "The knowing camera 2: recognizing and annotating places-of-interest in smartphone photos," in *Proceedings of International ACM SIGIR Conference on Research & Development in Information Retrieval*, pp. 707–716, 2014.

45. W. B. Lee, M. H. Lee, and I. K. Park, "Photorealistic 3d face modeling on a smartphone," in *Proceedings of IEEE Computer Society Conference on Computer Vision and Pattern Recognition Workshops (IEEE CSDL 2011)*, pp. 163–168, 2011.

46. H. Hakoda, W. Yamada, and H. Manabe, "Eye tracking using built-in camera for smartphone-based HMD," in *Proceedings of Adjunct Publication of the ACM Symposium*, pp. 15–16, 2017.

47. D. College, "Mobile sensing group." [Online], Available: http://sensorlab.cs.dartmouth.edu/.

48. M. Werner, M. Kessel, and C. Marouane, "Indoor positioning using smartphone camera," in *Proceedings of IEEE International Conference on Indoor Positioning and Indoor Navigation (IEEE IPIN 2011)*, pp. 1–6, 2011.

49. S. Kwon, H. Kim, and K. S. Park, "Validation of heart rate extraction using video imaging on a built-in camera system of a smartphone," in *Proceedings of International Conference of the IEEE Engineering in Medicine & Biology Society (IEEE EMBC 2012)*, pp. 2174–2177, 2012.

50. R. Boubezari, H. L. Minh, Z. Ghassemlooy, and A. Bouridane, "Smartphone camera based visible light communication," *Journal of Lightwave Technology*, vol. 34, no. 17, pp. 4121–4127, 2016.

51. E. Miluzzo, N. D. Lane, R. Peterson, H. Lu, M. Musolesi, S. B. Eisenman, X. Zheng, and A. T. Campbell, "Sensing meets mobile social networks: the design, implementation and evaluation of the cenceme application," in *Proceedings of ACM Conference on Embedded Network Sensor Systems (ACM SenSys 2008)*, pp. 337–350, 2008.

52. X. Liang and G. Wang, "A convolutional neural network for transportation mode detection based on smartphone platform," in *Proceedings of IEEE International Conference on Mobile Ad Hoc and Sensor Systems (IEEE MASS 2017)*, pp. 338–342, 2017.

53. R. K. Rana, C. T. Chou, S. S. Kanhere, N. Bulusu, and W. Hu, "Ear-phone: an end-to-end participatory urban noise mapping system," in *Proceedings of the 9th ACM/IEEE International Conference on Information Processing in Sensor Networks (ACM IPSN 2010)*, pp. 105–116, 2010.

54. R. Nandakumar, S. Gollakota, and N. Watson, "Contactless sleep apnea detection on smartphones," in *Proceedings of the 13th Annual International Conference on Mobile Systems, Applications, and Services (ACM MobiSys 2015)*, pp. 45–57, 2015.

55. S. Yun, Y.-C. Chen, and L. Qiu, "Turning a mobile device into a mouse in the air," in *Proceedings of the 13th Annual International Conference on Mobile Systems, Applications, and Services (ACM MobiSys 2015)*, pp. 15–29, 2015.

56. W. Mao, J. He, and L. Qiu, "Cat: high-precision acoustic motion tracking," in *Proceedings of the 22nd Annual International Conference on Mobile Computing and Networking (ACM MobiCom 2016)*, pp. 69–81, 2016.

57. W. Wang, A. X. Liu, and K. Sun, "Device-free gesture tracking using acoustic signals," in *Proceedings of the 22nd Annual International Conference on Mobile Computing and Networking (ACM MobiCom 2016)*, pp. 82–94, 2016.

58. R. Nandakumar, V. Iyer, D. Tan, and S. Gollakota, "Fingerio: Using active sonar for fine-grained finger tracking," in *Proceedings of the 2016 CHI Conference on Human Factors in Computing Systems (ACM CHI 2016)*, pp. 1515–1525, 2016.

59. S. Yun, Y.-C. Chen, H. Zheng, L. Qiu, and W. Mao, "Strata: Fine-grained acoustic-based device-free tracking," in *Proceedings of the 15th Annual International Conference on Mobile Systems, Applications, and Services (ACM Mobisys 2017)*, pp. 15–28, 2017.

60. Y. Xuan and B. Coifman, "Lane change maneuver detection from probe vehicle DGPS data," in *Proceedings of IEEE Intelligent Transportation Systems Conference (IEEE ITSC 2016)*, pp. 624–629, 2006.

61. F. Peyret, J. Laneurit, and D. Betaille, "A novel system using enhanced digital maps and WAAS for a lane level positioning," in *Proceedings of 15th World Congress on Intelligent Transport Systems and ITS America's 2008 Annual Meeting*, pp. 1–12, 2008.

62. novatel, "High precision GNSS receivers." [Online], Available: https://www.novatel.com/products/gnss-receivers/, 2016.

63. ublox, "Neo-m8p series." [Online], Available: https://www.u-blox.com/en/product/neo-m8p-series, 2016.

64. M. Rohani, D. Gingras, V. Vigneron, and D. Gruyer, "A new decentralized Bayesian approach for cooperative vehicle localization based on fusion of GPS and inter-vehicle distance measurements," in *Proceedings of International Conference on Connected Vehicles and Expo (IEEE ICCVE 2013)*, pp. 473–479, 2013.

65. L. C. Bento, R. Parafita, and U. Nunes, "Inter-vehicle sensor fusion for accurate vehicle localization supported by v2v and v2i communications," in *Proceedings of 15th International IEEE Conference on Intelligent Transportation Systems (IEEE ITSC 2012)*, pp. 907–914, 2012.

66. H. Li and F. Nashashibi, "Multi-vehicle cooperative localization using indirect vehicle-to-vehicle relative pose estimation," in *Proceedings of IEEE International Conference on Vehicular Electronics and Safety (IEEE ICVES 2012)*, pp. 267–272, 2012.

67. W. Hedgecock, M. Maroti, J. Sallai, P. Volgyesi, and A. Ledeczi, "High-accuracy differential tracking of low-cost GPS receivers," in *Proceeding of the 11th annual international conference on Mobile systems, applications, and services (ACM MobiSys 2013)*, pp. 221–234, 2013.

68. K. Golestan, S. Seifzadeh, M. Kamel, F. Karray, and F. Sattar, "Vehicle localization in VANETs using data fusion and V2V communication," in *Proceedings of the second ACM international symposium on Design and analysis of intelligent vehicular networks and applications (ACM DIVANet 2012)*, pp. 123–130, 2012.

69. V. Cevher, R. Chellappa, and J. H. Mcclellan, "Vehicle speed estimation using acoustic wave patterns," *IEEE Transactions on Signal Processing (IEEE TSP)*, vol. 57, no. 1, pp. 30–47, 2009.

70. V. Tyagi, S. Kalyanaraman, and R. Krishnapuram, "Vehicular traffic density state estimation based on cumulative road acoustics," *IEEE Transactions on Intelligent Transportation Systems (IEEE TITS)*, vol. 13, no. 3, pp. 1156–1166, 2012.

71. H. Li, H. Dong, L. Jia, D. Xu, and Y. Qin, "Some practical vehicle speed estimation methods by a single traffic magnetic sensor," in *Proceedings of 14th International IEEE Conference on Intelligent Transportation Systems (IEEE ITSC 2011)*, pp. 1566–1573, 2011.

72. B. Hoh, M. Gruteser, R. Herring, J. Ban, D. Work, J.-C. Herrera, A. M. Bayen, M. Annavaram, and Q. Jacobson, "Virtual trip lines for distributed privacy-preserving traffic monitoring," in *Proceedings of the 6th international conference on Mobile systems, applications, and services (ACM MobiSys 2008)*, pp. 15–28, 2008.

73. A. Thiagarajan, L. Ravindranath, K. LaCurts, S. Madden, H. Balakrishnan, S. Toledo, and J. Eriksson, "Vtrack: accurate, energy-aware road traffic delay estimation using mobile phones," in *Proceedings of the 7th ACM Conference on Embedded Networked Sensor Systems (ACM SenSys 2009)*, pp. 85–98, 2009.

74. G. Chandrasekaran, T. Vu, A. Varshavsky, M. Gruteser, R. P. Martin, J. Yang, and Y. Chen, "Tracking vehicular speed variations by warping mobile phone signal strengths," in *Proceedings of IEEE International Conference on Pervasive Computing and Communications (IEEE PerCom 2011)*, pp. 213–221, 2011.

75. G. Chandrasekaran, T. Vu, A. Varshavsky, M. Gruteser, R. P. Martin, J. Yang, and Y. Chen, "Vehicular speed estimation using received signal strength from mobile phones," in *Proceedings of the 12th ACM international conference on Ubiquitous computing (ACM UbiComp 2010)*, pp. 237–240, 2010.

76. D. Gundlegard and J. M. Karlsson, "Handover location accuracy for travel time estimation in GSM and UMTS," *Intelligent Transport Systems IET*, vol. 3, no. 1, pp. 87–94, 2009.

77. S. Thajchayapong, W. Pattara-atikom, N. Chadil, and C. Mitrpant, "Enhanced detection of road traffic congestion areas using cell dwell times," in *Proceedings of IEEE Intelligent Transportation Systems Conference (IEEE ITSC 2006)*, pp. 1084–1089, 2006.

78. M. Sakairi and M. Togami, "Use of water cluster detector for preventing drunk and drowsy driving," in *Proceedings of IEEE Sensors*, pp. 141–144, 2010.

79. D. Lee, S. Oh, S. Heo, and M. Hahn, "Drowsy driving detection based on the driver's head movement using infrared sensors," in *Proceedings of Second International Symposium on Universal Communication (IEEE IUCS 2008)*, pp. 231–236, 2008.

80. C.-W. You, N. D. Lane, F. Chen, R. Wang, Z. Chen, T. J. Bao, M. Montes-de Oca, Y. Cheng, M. Lin, L. Torresani, et al., "Carsafe app: Alerting drowsy and distracted drivers using dual cameras on smartphones," in *Proceeding of the 11th annual international conference on Mobile systems, applications, and services (ACM MobiSys 2013)*, pp. 13–26, 2013.

81. D. Chen, K.-T. Cho, S. Han, Z. Jin, and K. G. Shin, "Invisible sensing of vehicle steering with smartphones," in *Proceedings of the 13th Annual International Conference on Mobile Systems, Applications, and Services (ACM MobiSys 2015)*, pp. 1–13, 2015.

82. S. Lawoyin, X. Liu, D.-Y. Fei, and O. Bai, "Detection methods for a low-cost accelerometer-based approach for driver drowsiness detection," in *Proceedings of IEEE International Conference on Systems, Man and Cybernetics (IEEE SMC 2014)*, pp. 1636–1641, 2014.

83. C. Karatas, L. Liu, H. Li, J. Liu, Y. Wang, S. Tan, J. Yang, Y. Chen, M. Gruteser, and R. Martin, "Leveraging wearables for steering and driver tracking," in *Proceedings of IEEE International Conference on Computer Communications (IEEE INFOCOM 2016)*, pp. 1–9, 2016.